THE WORLD OF ANTS

THE WORLD OF ANTS
A Science-fiction Universe

RÉMY CHAUVIN

Translated by

GEORGE ORDISH

 HILL AND WANG NEW YORK

First published in French
© 1969 by Librairie Plon, 8, rue Garancière, Paris 6e
English translation © 1970 by George Ordish
All rights reserved
ISBN 0-8090-9810-5
Library of Congress Catalog Card Number 78-148236
First American edition March 1971
Manufactured in the United States of America

1 2 3 4 5 6 7 8 9 0

595.7

Ants — Behavior

CONTENTS

Instead of a preface . . .   9

   I The ant as a friend of man   26

  II Other, more or less likeable ants   70

 III The ants' herds   92

 IV City life   104

  V The vices and enemies of the city   128

 VI Social relations   143

VII The brain and senses of ants   163

Conclusion   195

Index   211

LIST OF PLATES

Between pages 32 and 33.

An example of the constant renewal of an ant-hill's surface.

An *Amorphocephalus* seeks food from a worker ant.

The "fear of the void" in red ants.

An *Atta* queen on a fungus bed.

Three workmen excavating an *Atta* nest.

The hunting column of *Anomma wilwerthi*.

Male *Anomma*.

Column of *Anomma wilwerthi* leaving the nest in a sunken track
worn down by movement of large numbers along it.

Living queen, *Anomma wilwerthi*.

Sunken exit track of *Anomma wilwerthi*.

Anomma wilwerthi, preliminaries to the formation of the
temporary nest.

INSTEAD OF A PREFACE . . .

WHERE DOES THE CLASSIFICATION OF ANTS COME FROM?

The anatomy and physiology of ant-lovers, "myrme-cologists", from the Greek

Myrmecologists are a species of the large genus "entomologists". A good classification first of all needs a definition of the genus, only proceeding later to the characteristics of the species.

An entomologist is not inevitably the harmless eccentric found in the school prize books of my childhood, with long hair, clothes flying and a big hat with a few butterflies or beetles pinned to it; often he was pictured contemplating rather abstractedly through a magnifying glass some minute insect, at the base of a flower corolla. Around this picture would be set the famous dictum *"Natura maxime miranda in minimis"* (which some writers put as *"In minimis maximus Deus"*.): Nature is most to be admired in the smallest. I can even remember the rest of the book: it was about an uncle who took his two nephews for a country walk describing at every step the insects they met. When I was about twelve the book charmed me, in fact until I found stronger meat—the *Souvenirs entomologiques* of the great Fabre. How many of us entomologists owe our vocation to this man?

And what is an entomologist's vocation? I think its essence lies in the same sort of astonishment as we experience on first seeing the sea: the sea of insects, the infinite variety of colours and shapes, the swarming mass of instincts among the thousands, tens of thousands, hundreds of thousands of species; just remember, for instance, that more than 600,000 species have been described and that there must be at least double that number in existence. True entomologists have a sort of fever that stays with them as long as they live. Some pass their lives in collecting and classifying these six-legged beings and can never tear themselves away from a fascinated contemplation of these creatures of Nature, so rich, so varied and even at times so crazy.

They would be easy to understand . . . but the myrmecologist branch diverges early from the family tree. Myrmecologists are born of a different amazement, their attention focuses on a different object, usually by the discovery, deep in the woods, of the City of Utter Otherness, the city of the ants, with its ceaseless movement, tiny yet effective.

On that day the potential myrmecologist sees a spectacle so strange (mankind's antipodes) that it might be happening on another planet. I imagine his astonishment must be even greater in the tropics when he excavates an *Atta* nest, as did Jacoby, and first sees tremendous architecture and the fungus gardens. Or yet again when he searches through the crammed storehouses of the harvesting ants of North Africa. Or perhaps— though unfortunately I only know of this by hearsay—when he has the good fortune to stumble upon a "super-colony" of a hundred or more nests, pressed one against the other over an area of several hectares and agitated in proportion.

That is how you become a myrmecologist, because you want to understand the ants' mysterious organization; and also because you realize all of a sudden that man *is not the only solution* evolution has achieved; perhaps not even the best, who knows? And who knows whether on other planets primates may never have progressed beyond the tree-living stage, whilst ants have led the way and have acquired intelligence? One day we shall find out.

But it is time to become acquainted with the world of ants, first answering two simple questions: How are they classified? How are they constructed?

Unfortunately on the way I shall have to force a number of Latin names on you, each one more outrageous than the last. It is not my fault, ants do not have any other names . . . it is the pill hidden in the jam of myrmecology.

Classification of ants

Like all systematic studies this is not exactly enjoyable, but we cannot very well begin this book without it. It is not an easy subject. Ants often resemble each other closely and it is only by rather minimal characteristics that we can distinguish species one from the other. Nevertheless, as can be seen by looking closely at Fig. 1, certain parts of the body, such as the head, show marked differences.

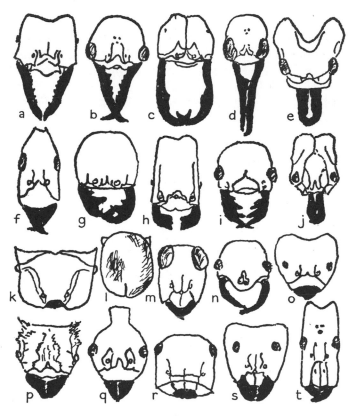

FIG. 1

Heads of a number of ants (after Wheeler, slightly modified):
(a) *Mystrium rogeri*; (b) *Myrmecia gulosa*; (c) *Eciton hamatum*, soldier; (d) *Harpegnathus cruentatus*, queen; (e) *Daceton armigerum*; (f) *Leptomyrmex erythrocephala*; (g) *Cheliomyrmex nortoni*, soldier; (h) *Pheidole lamia*; (i) *Thaumatomyrmex mutilatus*; (j) *Odontomachus haematodes*; (k) *Cryptocerus clypeatus*, soldier; (l) *Cryptocerus varians*, soldier; (m) *Opisthopsis respiciens*; (n) *Leptogenys maxillosus*; (o) *Azteca sericea*, soldier; (p) *Acromyrmex octospinosus*; (q) *Dolichoderus attelaboides*; (r) *Colobopsis impressa*, soldier; (s) *Camponotus cognatus*, soldier; (t) *C. mirabilis*, queen.

Some 6,000 species of ants are known at present and new ones are frequently found: they all have a more or less large head, a thorax joined to the abdomen by a *petiole* more or less long and thin, and a swollen abdomen, or *gaster*. Usually the colour is dark, black or brown, but some ants are bright red mingled with white, and on trees in the tropics one can even find green ants, though this latter colour is really very rare in ants. *Their habits, however, are enormously varied and it is by using these habits to assist morphology* that we can separate the primitive ants from the more developed ones. We will start with the most primitive.

The *Ponerinae* are long ants, not very numerous in our climate, but plentiful in the tropics. They include a number of giants, such as *Megaponera* and *Paltothyreus* of Africa (see Fig. 2).

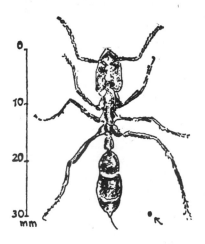

FIG. 2

A giant of the ant world: the Argentine *Dinoponera*. The dot at the bottom right is a minor worker of *Pheidole* on the same scale (after Gœtsch).

Many of these insects are 2 cm. or more in length and nature has seen to it that they are well armed, either by a sting, often a veritable sword, which they know how to use, or even by an insufferable smell, as, for instance, the corpse ants or *Palto-thyreus*. I remember crushing one by accident one day walking along the shore of Abidjan lagoon. I had to decamp! The faecal smell is comparable to that of a stink-bomb, but much stronger: the workers only squirt it out when they are uneasy or wounded. These ants are great hunters of termites, and they are led by a scout who has found the hunting ground beforehand and who is indispensable to them. If this scout is removed the black band disperses or goes straight back to the nest. It is there we find the champion territorial expansionists of the ant world. They all have hunting territory where they object to intruders, and this territory can vary from a few square metres to a hectare per nest in the case of our red ants. However, the enormous bull-dog ant of Australia, *Myrmecia gulosa*, 2 cm. long, wanders over several hectares, will attack at several metres' distance an entomologist watching it and chase him as he flees. As it not only bites but also wounds with its huge sting it is rightly feared by Australians. Fortunately, its nests are not very populous, a few thousand individuals at the most, otherwise the areas it inhabits would be quite untenable by man. Think what our forests would become if our red ants, with their enormous population, were a centimetre larger.

Population explosion

The *Dorylinae* are the hunter ants, the famous driver ants of Africa; we have much to say about them on page 71. However, in passing we may note a giant queen of the *Anomma*, more than six cm. long, who, at the same time, holds the ant-egg laying record: she can in fact lay 60,000 eggs in three days and the next brood would not, according to Raignier, fit very readily into three ordinary buckets. Raignier also states that a colony can exceed 20 million individuals!

The *Promyrmicides* all belong to warm regions and are remarkable above all for their larvae, which have series of flattened extensions along their bodies which, no doubt, are licked by the workers. The genus *Pseudomyrma* lives on myrmecophytes, that is on ant-plants: trees or bushes having swellings or spines which are hollow or full of tender cells the workers can easily remove. The whole colony shelters there, and certain species are found only in such places; moreover, they seem to do no damage to the plant.

The *Myrmicinae* are possibly the most numerous family of ants and include our red ant with its painful sting. They are small or even minute at times. The *Leptothorax* nest in the crutch of a rotten branch, and the whole nest, with queen and brood, could easily be held on a thumbnail. They are seldom seen because they are nocturnal. However, I do know some woods near Paris where one can easily find three or four nests to the square metre.

The African *Crematogaster*, with a very painful sting, make nests of a paper-like substance, attached to trees. Often enormous nests have a well defined structure, a fairly rare thing with ants.

The famous *Messor* harvesters are found among the *Myrmicinae*; they store many kinds of grain, but they are not the only ants to do this for many very different species do the same. Bernard says they cut down a plant at its base in order to harvest the grain more easily. They are veritable harvesting champions and it has been estimated that in Algeria they steal a tenth or more of the cereal crop. Some nests are 50 metres wide and 3 metres deep with thousands of storage cavities.

Burrowing champions

The *Atta* fungus growers, dealt with on page 79, are also avid harvesters: often, over a few seasons, a nest can collect up to *five tons* of leaf, used for growing fungi. The nest itself has been made by excavating several tons of soil, and it is among these ants that the biggest cavities made by an insect can be found. If need be a man could get into one, yet we do not really know to what use the ants put these great hollows.

This family contains the only known ant enemy of termites, *Paedalgus termitolestes*, actually within the nest. The almost blind workers of this ant rob termite nests: the sexual forms of this ant species are relatively enormous.

There is nothing particularly remarkable about the *Dolichoderinae*; some of them make nests of paper, or live upon myrmecophytes. One species (*Aneuretus*) lives in the cup of insect-catching carnivorous plants and seems to be perfectly happy there (see page 107).

Harvesting and hunting champions

The *Formicinae*. Unlike most of the species mentioned up to now the *Formicinae* inhabit the temperate regions. They have no sting but can, like our red ants, spray their enemies with a notable quantity of poison containing a high concentration of formic acid, which makes them no less formidable. Some members of this family are among the most remarkable as regards their behaviour, although they neither gather cereals nor collect fungi: for it is with them that *stock raising has reached its highest peak* (see page 92). They are also *great hunters and great collectors of honeydew from aphids.* A large nest of red ants will in fact need a hundred kilos of honeydew a year and a kilo of insect meat per day.

Camponotus is the largest ant in the European region. We should also notice the *Colobopsis*, where a large worker with a flat head carefully seals the entrance to a nest and allows only its companions to enter. The famous *Œcophylla*, or weaver ants, stitch several leaves together with silk thread, up to the size of a handkerchief at times. Finally, the family contains *slavemakers. Formica sanguinea* and *Polyergus rufescens* find it convenient to steal nymphs from other species; when these nymphs become adult they quite willingly work for their masters, although of a different species.

The champions of size difference

Differences in size within a species (polymorphism) can be more or less marked between different workers, or between them and the sexuals. I think the champion of size difference between workers must be *Pheidologeton diversus*, of Indonesia: *thirty of the smaller workers (minores) can be accommodated on the head of one of the larger workers (majores)*! As to differences between workers and sexuals, these can be slight, as in the case of our red ants, or almost non-existent, as in *Myrmica*: it can even happen (though rarely) that the queen is smaller than some of the large workers. On the other hand, I believe the biggest difference is certainly found among the magnans or driver ants, with their enormous wingless queen *several thousand times* bigger than the *minor* workers (see Figs. 3 and 4).

Physiology

There is not much to say on this: the ant is a typical insect, with a simple, straight digestive gut, into which the secretory organs, or Malpighean tubes, open. A cardiac vessel runs along the upper part of the abdomen with numerous contracting chambers which "agitate" the blood rather than really pumping

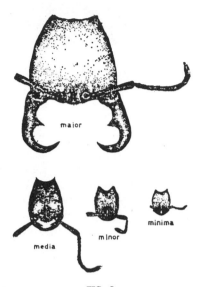

Differences in the heads of *Anomma wilwerthi* workers (after Boven, 1957).

it. Moreover this blood has no respiratory function, any more than with other insects. It serves to circulate food particles and waste products for excretion. Breathing and oxygenation of the tissues is secured by tubes full of air, the tracheae, which convey oxygen to the sites in all the organs where it is needed. In short, no special difference from any other insect.

However, ants, all of which are alike in this respect, do have *some special anatomical features* which distinguish them from all other Hymenoptera. These are the *infrabuccal sac*, the *metanotal gland* and the *proventricular valve* of the *anterior intestine*. It has been suggested that these are necessary for a social life; possibly this is so, but it has not been proved. The *proventricular gland* might in fact well suit a fundamental action common to all social insects, namely the exchange of food, or trophallaxis (see p. 145). As to the metanotal gland at the back of the thorax, we do not know the purpose of its secretion, except that it does produce the stinking liquid of *Paltothyreus*, and, moreover, we are not very much better informed on the many glands ants possess (see below p. 19). Regarding the infrabuccal sac, this

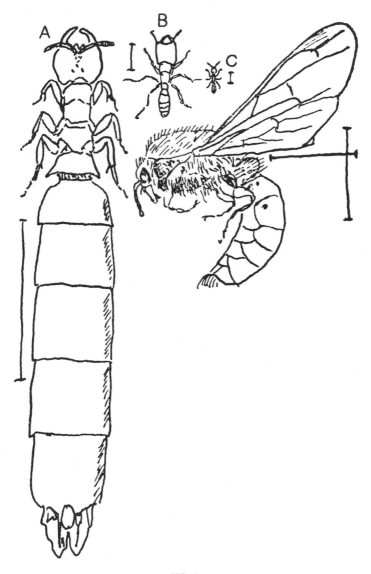

FIG. 4

Castes of *Dorylus helvolus* drawn to the same scale (after Emery):
A. Female; B. *Major* worker; C. *Minor* worker; *below*, male.

collects the debris that an ant collects on its body, or on those of its colleagues, by means of the endless mutual licking on which they spend almost as much time as they do on trophallaxis (food-exchange). It may also serve as a filter since, if

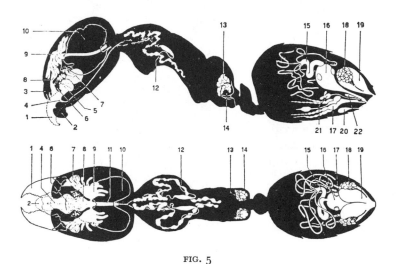

FIG. 5

Diagram showing glands and some of an ant's internal organs (after Pavan, 1957). *Above*, vertical section. *Below*, horizontal section. 1. Mandibles; 2 and 3. Mouth parts; 4. Opening of labial glands; 5. Infrabuccal sac; 6. Prepharyngeal plate with the orifices of the maxillary glands; 7. Mandibular glands; 8. Pharynx; 9. Pharyngeal gland; 10. Brain; 11. Oesophagus; 12. Joining of the anterior and posterior branches of the labial glands; 13. and 14. Metanotal glands; 15. Anterior part of the proctodaeum with the Malpighian tubes; 16. Rectum sac with rectal glands; 17. Ovary and last abdominal ganglion; 18. Anal glands; 19. Storage sac for 18; 20. Ventral organ; 21. Venom acid gland; 22. Alkaline gland.

one mixes fine mineral particles with liquid food, the minerals are found in the little pellets that the ants leave in the nest, and which are none other than the contents of the infrabuccal sac. The amount of filtering depends on the social life led, for it is much more prevalent in colonies than in isolated ants. There are two cases of special adaptations of the sac to particular conditions of the nest. *Atta* queens who are going to found a new colony always carry in it an abundant mycelium of the family's fungus. It has not necessarily been collected by the queen

"intentionally", for it is certain that in licking herself or anything else in the colony she cannot fail to pick up traces of the
fungus. What is certain is that in every case she spreads the
contents of her buccal sac over the first eggs she lays, then
crushes them, thus establishing the first culture bed for the
fungus. With the *Pseudomyrmecinae* the infrabuccal sac serves as
a mould and forms the food into pellets. These pellets are then
distributed to the larvae who take them into a special buccal
receptacle, which biologists have not been able to resist calling
by the barbarous name of the "trophothylax".

The glands of ants

The cephalic and cephalothoracic glands of ants have interested many writers, because they are easily seen and show many
variations over the course of the year. Among these is the
postpharyngeal gland, composed of a very voluminous bi-lobed
sac, which is not much developed in the young ant but reaches
its maximum size at the time of raising the sexuals. Moreover,
different ants show big differences in this respect, not related
to their functions or to whether their special work lies inside or
outside the nest. Bees have no analogous gland. The *maxillary
gland* is formed by two sacs opening into the mouth cavity.
Very similar glands are found in bees, which glands secrete the
famous *royal jelly*, the food of the sexual larvae and of queens.
The function of these glands in ants is not known; they are a little
larger in ants working inside the nest. In ants, the gland
secreting the sexuals' food is the enormous labial gland which
extends quite far into the thorax. Its activity increases forty
days after hatching, but its stage of development varies greatly
over the year; it is retracted in the case of the winter workers,
but a week's warmth is enough to re-activate it. I have kept the
mandibular glands till last, they are the same size in all individuals, as is the similar and like-named gland in bees: I
deal with this below (page 160). It appears to secrete substances
notifying alarm or recognition.

The poison gland has a very special structure: it consists of a
pair of filaments floating in the body cavity that trail behind
them a long, coiled tube, which, unwound, would be 20 cm.
long. It is enclosed in the wall of the poison vesicle (for poisons
see page 161). I cannot resist quoting some startling figures here:

in the course of a fight a worker may discharge 1 mg. of poison, which is equal to 0·5 mg. of formic acid. Suppose you try the experiment my assistant and I carried out in the Vosges mountains. We mercilessly stirred up an enormous nest, bigger than both of us; like us you would have to retreat half asphyxiated by formic acid. If we suppose that the nest had only 50,000 ants spraying us, which is nothing, that is equal to 25 g. of formic acid. Much less would be enough to stop the respiration of several men. Another curious fact: formic acid is highly insecticidal, as has been known for a long time. But the poison of another species, the fire ant *Solenopsis saevissima*, also kills by contact (even though the ant has a sting) and indeed is almost as toxic as DDT; in addition it has antibiotic properties.

Some special diets

In general ants use anything for food, particularly the hunters who eat anything they can catch (see page 73). There is, however, *some specialization*. For instance, the Dacetines only hunt two kinds of collembola; *Myopias* attacks only millepedes and *Leptogenys* only woodlice. A large number of ants show an almost exclusive appetite for termites. As to sugary diets, they do not all show the eclecticism of the red ants, who milk more than sixty species of aphids; but few species confine themselves to only two or three kinds of aphid and scale insects.

Swarming of the sexuals

Copulation is carried out much as in the case of bees, in that the males fly in swarms, sometimes enormous, over well-defined areas. For instance, with *Myrmica* it is over a bare and flat area, near rising ground or a wall. The males persistently return to this area even if a gust of wind blows them out of it. It is the females who come to these swarms. We do not yet know if, as with bees, these male swarms keep the same swarming place generation after generation, for twenty years or more.

The females usually mate with several males; the huge *Atta* and *Anomma* queens need several males to fill their seminal vessels, which have to suffice for several years for the procreation of millions of workers. In the case of the parasitic ants, such as *Anergates atratulus*, a parasite of *Tetramorium caespitum*, the male not only has no wings, but can hardly even walk. Thus he

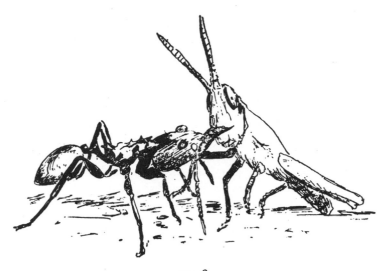

FIG. 6
Daceton armigerum capturing a small cricket (after Roo).

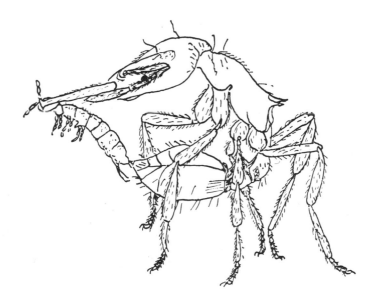

FIG. 7
Worker of *Strumigenys ludia* seizing and stinging a small collembola
(after Brown and Wilson, 1959).

mates with his sisters inside the host's nest. Both winged and wingless males are found in the species *Ponera eduardi*; the latter mate inside the nest and always with workers, never with the queen. These workers lay eggs which do not develop. It is true that these observations have only been made on laboratory cultures and we do not know what happens under natural conditions. We are also ignorant as to how the enormous *Anomma* male finds its queen, bigger still and wingless, guarded by an army of female workers. The male must somehow get itself accepted by the colony, and some observers have seen males with torn wings moving around the heart of the nest.

When the queen has been fertilized she twists in all directions, and begins a strange series of movements tending to remove her wings. The great Belgian myrmecologist, Raignier, has described the scene very strikingly in the case of a *Camponotus* queen. "The queen travels round her new place of residence two or three times, where she seems to be quite at ease. She stops, cleans her antennae and feet, then suddenly, with a blow from the knee of her right middle leg, pushes the right forewing towards the front of her body. The animal leans slightly to the left whilst the middle and hind legs become very active, working feverishly to detach the wing. After two minutes this gives way and falls to the ground. The queen now rapidly runs round again, stops and goes through the same motions to release the right hind wing and soon it, too, comes off. The auto-amputation then continues on the left side. The queen straightens out her body and leans to the right. The left-hand legs now carry out identical movements. The two left wings are detached in turn and fall to the ground. Once more she carefully grooms herself. . . ." All ant queens behave in this way, except the enormous, blind, wingless queens of *Anomma*. After the wings have gone the big flight-muscles are reabsorbed into the insect's body, becoming available for the general metabolism of the insect. It is thanks to this that the first eggs are formed; and this is why the queen can live and lay in spite of a sometimes very lengthy fast in certain species (up to a year in some cases).

What are the ant's origins?

Ants are much older than man, a hundred times older or more. It was thought that they had been found even in the

Secondary era in the lower lias 200 million years ago. The fossils Heer found in 1865 were named by him *Paleomyrmex prodromus*, undoubtedly an error, for the insect so named was not even Hymenopterous. Nevertheless the large number of fossil ants in the Tertiary epoch points to the existence of a source in the Secondary, though there is but little exact information on this up to the present.

In fact, in the Tertiary, and more exactly at its start, the Eocene and Oligocene strata of about 80 million years ago, large numbers of ants are found in America and Europe in all sorts of deposits, particularly in the celebrated Baltic amber: this is fossilized pine resin, which then, as now, captures a crowd of insects, particularly ants, because of its sticky nature. The creatures in this solid and antiseptic medium are in an unbelievably good state of preservation; in some cases you can dissolve the fossil resin and mount the ant on a slip of cardboard just as you do insects today, and often nothing would distinguish the fossil from present-day species, were it not for its respectable age.

I write that nothing can distinguish it because we find some species from that far-off period which are just the same as existing ones, for example *Formica fusca*. Moreover, ants must have been numerous, because, of 600 fossil species of Hymenoptera found, 307—more than half—are ants. Specimens are found in great numbers; for instance, Wheeler examined more than 5,000; 139 specimens were certainly *Camponotinae*, 25 *Dolichoderinae*, 185 *Myrmicinae* and 27 *Ponerinae*. One specimen was a *Dorylinae*. A few scarce genera are extinct, but we are never quite sure of this, for a live *Stereomyrmex*—a genus hitherto known only as a fossil—was found recently in Ceylon. It is not so very unusual to find species of higher animals and plants alive and well that have been thought extinct for hundreds of millennia. *F. fusca* is not the only living ant also found fossilized in amber; *Lasius niger* and *Ponera coarctata* are also discovered there, and the most exacting morphologist would not be able to distinguish them from present-day specimens.

Another strange thing, as Wheeler remarked, is the *absence of polymorphism* among the Tertiary fossil ants: some of the same species existing today (*Dimorphomyrmex, Camponotus*) show a marked polymorphism. Obviously one could object that an absence proves nothing and one could postulate that the soldiers, if they existed, did not go any further to forage than do those of today, and consequently ran less chance of being fossilized in resin, for example. To this the reply is that a little later (twenty or thirty million years later, but what is that when dealing with geological epochs?) polymorphism becomes very obvious among fossil ants.

Remember that fossils of the pre-human beings are scarcely older than a million years; as to *Homo sapiens,* he can scarcely be put back more than 100,000 years, a thousand centuries. This is insignificant compared with the *eight hundred thousand centuries* of the ants; we were born yesterday.

The origins of social insects

This is a very speculative field, nevertheless one cannot refrain from asking the question: what is the origin of social habits in insects? Were they social from the start? Or rather, as they are obviously the peak of some evolutionary process, can we find some non-social ancestors?

Entomologists and palaeontologists more or less agree that they are descended from parasitic Hymenoptera, because such parasites display, in embryo, many characteristics of behaviour strongly reminiscent of insect societies. For instance, eggs and larvae of these parasitic Hymenoptera have the same appearance as in ants, including the septum between the larva's mid and posterior intestines, which does not allow the faeces to be expelled until the last moult, thus helping to keep the nest clean. This closure is found in the larvae of parasitic Hymenoptera, but not in the other branch of this family, the Symphyta, vegetarians who branched off from the phylum of parasitic and of social Hymenoptera at an early stage.

Parasitic Hymenoptera already possess the faculty of pre-determining the number and sex of the eggs they lay according to the size and other characteristics of the host. The life of the female, moreover, is lengthened by what Flanders calls ovisorbtion, or the reabsorbing into the maternal tissues of unused eggs, the food reserves acquired from them thus being recycled. In this way with some parasites the longevity of the female exceeds the length of several generations of her offspring.

At the start of the evolution of ants all the females were equal and all could undertake the same tasks; there was but a simple tendency for all to live in the same nest. This is still found in certain genera of wasps or of the less highly evolved bees (*Belonogaster, Allodape*). The next stage evidently is the differentiation of the females into two castes, those capable of reproduction (queens) and the others (workers). Now the females of parasitic Hymenoptera are sometimes fecund and sometimes quite incapable of reproduction, owing to external causes (lack of suitable food or weather conditions). This state is still quite reversible.

We cannot readily account for the transition to another essential stage, that of *morphological differentiation*. It is true that among social insects themselves we find every transitional stage between the two extremes, queens hardly different from workers, who may even lay eggs in the presence of the queen, and the complete inhibition by the queen of all egg-laying by workers, with big morphological differences.

CHAPTER I

THE ANT AS A FRIEND OF MAN

Utility of red ants — Their nest — Their heating —
Food problems

Nᴏʙᴏᴅʏ ᴡᴏᴜʟᴅ ʜᴀᴠᴇ thought a few years ago that the red ants, those unpleasant inhabitants of our woods, could turn into allies for the protection of our forests. Yet such is the case: the red ant has now taken its place among the small band of insects useful to man.

Insects useful to man

It is strange what little use man has made of insects up to the present. They are, in fact, the only living creatures who oppose him, not without some successes. Think for a moment of the large mammals: they only exist because we want them to. If we cared to extinguish the race of elephants or whales over the face of the earth, we could do so with the greatest of ease. How different is the case of the mosquito!

On the other hand it seems that man has been very slow up to the present in his use of the six-legged world. The silkworm, the scales giving shellac, the bee, and you have the whole list. It is only comparatively recently that we have undertaken the breeding on an industrial scale of insects parasitic on other insects, with the object of securing biological control of pests, a process remarkably successful at times. To this must be added another use, which Bodenheimer has pointed out in a most interesting book, their use as food. For a long time it was asked how masses of people, whose diet was obviously deficient, managed to live on food which, as far as the investigators knew or saw, consisted of nothing but cereal porridge: later ethnologists, rather more observant, noticed that the women ceaselessly collected insects and made a kind of flour from them to add to the diet. The addition of vitamins and protein thus collected sufficed, and amply, to cover their basic needs.

But ants? Well, up to recently, no use had been found for them and in fact they were regarded as the most implacable enemies of man. Whilst the bee from the depths of time has been man's companion (but not his friend, for the bee has never been tamed and is quite unaware of the existence of the bee-keeper), whilst termites are the *pièce de résistance* and one of the richest protein sources for the very numerous termite-eating peoples, the ant has never been used, although, probably, certain tribes ate the grubs, which have quite an agreeable taste (moreover, bee larvae too are eaten quite often and are indeed a delicacy for the Japanese, recently even being preserved in tins!). In Europe these same larvae have been used as pheasant food. As collecting ant nymphs is not a very well-known trade it might be interesting to describe, after Gösswald, the technique of ant hunting.

Collecting nymphs

In spring and summer, in the early morning hours when the sun strikes the ant-heap, the ants mass their larvae and nymphs in the warmest spot. The ant-hunter quickly scoops up this portion of the heap and puts it in a sack. Previously he has prepared five pits, 25 cm. in diameter and 30 cm. deep, in bare ground, forming a circle about 2 metres in diameter: the pits are filled with fir or pine needles. The contents of the sack are turned out in a thin layer in the centre of the circle. The ants quickly set to work to carry the nymphs to apparent safety in the holes. When the ants have finished this the contents of the holes are put into the sacks once more, and the needles, with many ants on them, are easily pulled out; the nymphs sink to the bottom of the sack.

Two cleaning operations are all that remain to be done: the first is a winnowing, in which the ants are made to fall slowly from a height of about 1·6 metres. The nymphs, being heavier, fall almost vertically and the ants are blown further off by the wind. Finally, a piece of black muslin is passed several times over the heap of nymphs and the last remaining workers cling to the cloth and are removed.

The yield is at least one or two litres of nymphs per colony, with a maximum of three litres; but some operators, working on giant nests, which can be more than a cubic metre in size, may get 16 litres a year from it, that is more than 300,000 nymphs. Gösswald mentions that Styrian peasants used to put 50 to 70 hectolitres of nymphs on the market each year, which is equal to more than 130 million ants.

Another less-known method of exploitation is to use the needles and twigs on the ant-heaps as litter for cattle, thus replacing straw: peasants in Alsace and Germany collect them by the cartload.

Modern times

But all the above belongs in some degree to folklore. Gösswald had another idea born of an observation which must have been made thousands of times (in any case it was published in the eighteenth century), but never understood: the eternal history of science! When defoliating caterpillars attack a forest it usually suffers greatly, and the trees, having lost a large part of their leaves, take on a wintery appearance. *Except around the ants' nests where "green islands", greater or smaller in size, can be seen.* Gösswald supposed that perhaps this was due to the ants having eaten the caterpillars. And this was indeed the case.

From that discovery to the idea of breeding red ants artificially and then sending them to places where they were lacking was a big step. Nevertheless the German biologist accomplished it. However, before talking of "formiculture", we must go back a little to discuss the general biology of red ants, since it is well known today.

Species of red ants

The entomologists' first concern was to find out exactly what they were dealing with, and that is where they made an interesting discovery of which, in my opinion, not enough advantage has been taken.

In two words, what we call *red ants* in all probability includes several species, but we do not really know how to distinguish them. There seems to be one species called *Formica rufa*, hairier than the others, bigger, making bigger ant-heaps and with a not very large population, some 100,000 workers and

one queen per nest. At the other end of the scale is a smaller
species, sometimes called *rufo-pratensis* and sometimes *polyctena*,
less hairy, much given to burrowing and making slightly
raised heaps around which the earth is deeply worked, and
where each nest might have a million workers and 5,000 (I
repeat five thousand) queens. But we must also consider the
centre part of this chain, of which, up to the present, we only
know the two extremities. That is where things get complex,
because we find a number of species with a greater or smaller
number of queens—(they are called monogynous, oligogynous or
polygynous according to whether they have a single queen,
several queens or several thousand queens)—their hairiness and
morphology being uncertain. This is without counting that
Gösswald found within the species *polyctena* one race living in
pine-woods and another in fir stands, races distinguished by the
general appearance of the nest and by certain characteristic
behaviour. We may add that *rufa* lives happily in the underwood
whilst *polyctena* prefers sunny clearings.

Finally, and this helps to destroy one of the last criteria for
distinguishing species, many of these species *interbreed*, as has
been proved by Gösswald's followers. This is a situation to drive
systematists to despair, those people who love exact classification
and well-defined, dichotomic keys; however, in my opinion,
they ought to be pleased. The *rufa* case is not unique; similar confusion can be found in other cases, for example, with certain
mosquitoes and and with more than one species of hymenopterous parasite. It is as if, by chance, we found ourselves actually
looking at the boiling biological cauldron where the species are
being made. It is as if they were being created in front of our
very eyes. Because, as M. de la Palisse would say, before being
distinct species they were perhaps indistinct. The observation
ought not to depress us; on the contrary, everyone ought to be
very interested in it.

Biological data

However that may be, all the groups of red ants have the
same biology, which can now be outlined in broad terms. First
of all the *nests* are, as everyone knows, prominent heaps of
twigs and pine needles, and I shall have occasion to write of
their construction later on. First we must note a curious fact:

there is no relationship between the size of the nest and the size of the colony that lives in it. Furthermore, in the case of *polyctena*, colonies have a tendency to amalgamate into enormous super-colonies, derived from the joining of one or several mother colonies in permanent contact. One of these giant colonies, noted by Stammer in 1938, had 58 principal nests and 31 secondary ones, and the total length of the galleries connecting them was 8 kilometres. Father Raignier, the famous Belgian myrmecologist, reports on another of these super-colonies which had more than a hundred nests and covered about ten hectares. Very little is known with any exactitude about these phenomena of "two-tier" organization; I have often thought they could be one of the most difficult and interesting features in the biology of social insects. Who knows what interesting behaviour complexes might be found in these super-organizations? (See p. 206.)

According to Eckstein an isolated nest's hunting territory is about 3,500 square metres; but some will cover half a hectare. It varies greatly according to the nature of the ground, the strength of the colony and other circumstances.

Nuptial Flight and Mating

According to Marikovsky the nuptial flight of the red ant is preceded by the enlargement of the channels leading to the exits of the nest. The females come out, fly a short distance and rest on the lower branches nearby to await the males. Both females and males can copulate several times and fairly fierce fights can be seen between several males all wanting the same female. Marikovsky made some further observations in the Tien Shan mountains in Siberia. He noted vast assemblies of males and females; immediately after copulation the female, with a snap of her jaws, cuts off the male's abdomen from the thorax. The mutilated male flies away and takes several hours to die; the severed abdomen rapidly falls away from the female. Such cruel *amours* are by no means the speciality of ants, for many other similar instances are found among insects. The classic case is that of the female praying mantis who eats the male's head during copulation. Rœder showed it was essential for the male to lose his head in order that copulation could take place, for the brain exercised an inhibiting action on the process.

As nature always goes straight to the point, it was the only thing to do.

Once mated, the females of these red ant species cannot themselves found colonies; they are obliged to get into an established nest and reinforce the queens already there. Often they are destroyed in great numbers by the workers of their own or similar species. And very often even the workers who eventually will accept them show an aggressive attitude until one of them offers food to the strange female! This signifies that she has been adopted. It appears that young colonies, short of queens, more readily accept the new queens. Moreover, at the time of swarming not only do the young queens seek to enter a nest, but the old queens try to get out and perch on a branch to be fecundated again. Sometimes they are successful. But generally the workers strive to make the old queens re-enter the nest; their last resort is to introduce a droplet of formic acid between the mandibles of the queen, who then inclines her head and regains her home.

These Siberian ants seem to be very temperamental: things appear to be much calmer in our old Europe. In any case Gösswald has not noted all these phenomena in the course of his work on ant behaviour.

"Formiculture"

There are at least two methods of increasing ant colonies. The first consists simply in taking, at the beginning of spring, the ants and their twigs in a sufficient amount (some 20 litres). It must be done when a seething mass of ants first appears on the surface of the nest to warm itself in the sun. Often there is enough to fill several basins. One is thus sure to include queens in the consignment, since this is the only time during the year when they come out from the depths of the nest.

The ants and twiglets are then carried to the chosen spot, which should not be too shady nor too overgrown with bush. It sometimes happens that the ants thus transplanted will get up and march off some 50 metres; but if you have been careful to space out the transplants this will not matter much. There is not much chance of success with less than 15 to 20 litres of ants and twigs.

That is only the first method; another consists in using the

enormous surplus of queens usually lost each year, those who
do not get accepted by a nest or who fall to predators drawn to
the spot in great numbers by the bonus of basking ants. . . .
To do this Gösswald covers a nest in spring with a cone of dark
cloth, arranging a transparent receptacle at the top. The young
sexuals are attracted to the light and at first all rise to it, but
later fall back into a vessel put below it. Some nests produce
nearly all females and some only males. Several thousands of
each can be collected in an aquarium tank, the bottom of which
has been covered with moist earth and a few bits of bark; it is
then covered with fine metal mesh. The tank should be partly
in the sun, which encourages copulation. To enrich an existing
colony with these queens one must first give them the colony
odour: a few hundred workers are taken from the nest and
added to the tank with the new queens; a few days later a few
hundred more are added. They thus gradually give their smell
to the queens and at the end of two or three days they are all
let out near the nest, where they will be admitted without trouble.

Enemies of ants

Once an ants' nest has been established it seems to be
impregnable. The shots of formic acid are a decisive weapon;
no creature, vertebrate or invertebrate, would join battle with
the ants. If this was always the case they would have covered the
surface of the earth. They are, however, vulnerable at a
certain season, in winter, and in the sexual form. There are,
moreover, some insects that attack even the isolated workers.

There is no doubt that winter is the time the red ants suffer
most from predators. At the start of winter a number of holes of
varying depths may be seen around an ant-heap. These are
caused by woodpeckers, foxes and badgers. Woodpeckers are
capable of digging into the heart of the nest, 60 cm. below
ground or more, to feed on the torpid ants who are in no state to
defend themselves. Foxes and badgers do the same thing; they
are not after ants, but the fat larvae of a shining beetle, the
golden *Cetonia*, which live in the mould of the ant-heaps;
thanks to an abundant microbial flora in their intestines they
are able, by means of a miraculous bit of chemistry, to draw
nourishment from rotten wood exhausted by a thousand
fermentations, and grow big and fat from it. The ants do not

An example of the constant renewal of an anthill's surface. *Top:* three concentric polygons made with matches. *Middle and bottom:* the successive disturbance of the polygons. Photos at 3–4 hour intervals (after Chauvin).

Left: an *Amorphocephalus* seeks food from a worker ant (after Torassian).

Top, opposite: an *Atta* queen on a fungus bed, surrounded by *minor* and *minima* workers and one *major* worker, top centre.

Left: the "fear of the void" in red ants. A cardboard cylinder put vertically on a nest is rapidly filled with twiglets (after Chauvin).

Bottom, opposite: three workmen excavating an *Atta* nest. Note the enormous cavity in the middle, the exact purpose of which is not properly known (after Weber).

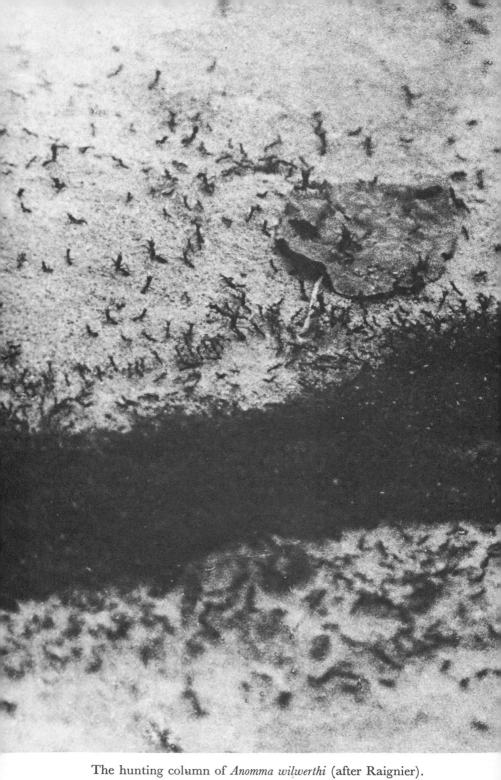

The hunting column of *Anomma wilwerthi* (after Raignier).

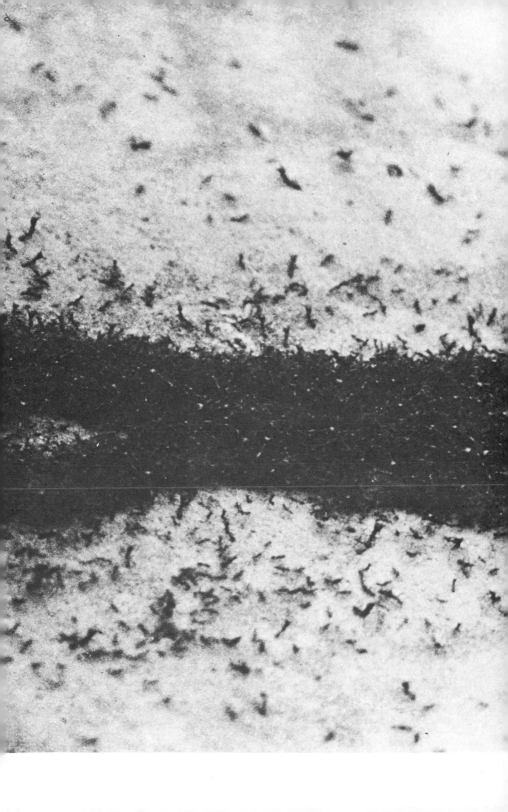

Male *Anomma*, 4 cm. long. These are the ants that swarm in such numbers round the lamps in tropical Africa (after Raignier).

Column of *Anomma wilwerthi* leaving the nest in a sunken track worn down by movement of large numbers along it (after Raignier).

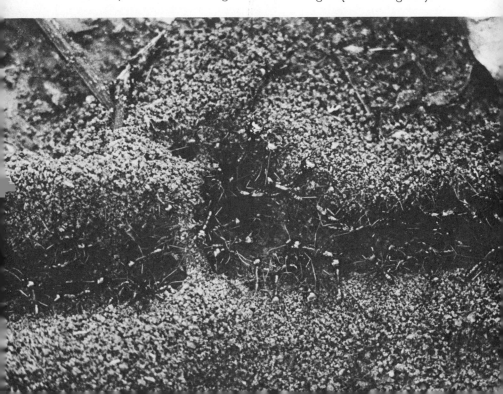

Living queen, *Anomma wilwerthi* (after Raignier).

Sunken exit track of *Anomma wilwerthi* (after Raignier).

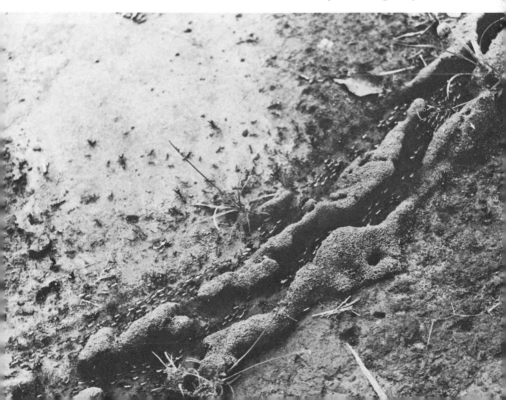

Anomma wilwerthi, preliminaries to the formation of the temporary nest ('bivouac'): garlands of ants hanging from a branch to the soil (after Raignier).

THE ANT AS A FRIEND OF MAN

attack them, perhaps because their skin is too smooth and resistant, or possibly because of a repellent secretion, as some observers maintain. But foxes and badgers will do anything to get them; their digging opens the nest and exposes the workers to their natural enemies.

The worst enemies of ants are certain other insects. For instance, the larvae of the beetle *Clythra*. The female beetle lets her eggs, covered with a brown substance that makes them look like bits of wood, fall on to the nest. The larvae from these eggs, sheltered by a hard case the ants cannot penetrate, feed on the eggs, larvae and pupae of their hosts and retire to their shelter at the first sign of trouble.* It is true that far-seeing nature does not allow *Clythra* to attack ants without some check: the predator is itself attacked by a *Mutillid, Smicromyrme montana,* which looks very like an ant. Adult *Clythra* fly very well, come out in the spring and are often captured at the same time as winged ants when these last are being collected for "formiculture".

Another insect, a Hymenopteron, *Elasmosoma*, attacks adult ants. The female, quick as a flash, places her egg on the worker who seems to suspect the danger and tries to reach the parasite. The same thing happens with other ants, the fungus-eating *Atta*, but in this case a small worker chases the parasite and, if the latter succeeds in putting its egg on a large worker, removes it. The red ants do not know how to help each other in this way. The *Elasmosoma* larva eats into the ant's abdomen. The worker ant, when about to die, leaves the nest and clings to a piece of grass. The parasite then leaves its host and pupates in the surface layer of soil.

Adult ants can also be attacked by worms, *Mermis*, which exhaust them and by mites (as are bees, but it is not the same species of mite). There is also a guest, frequently found among red ants, the minute ant *Formicoxenus nitidulus*, which seems to be sub-social; it lives in small groups of about a hundred individuals in the empty shells left by *Clythra* larvae when they have departed to become adult. It does not seem to harm the workers; perhaps it even obtains food from them from time to time.

---

* It often happens that the ant puts eggs into the passage made by the withdrawal of the *Clythra* larva, which then only has to turn round and feast on them.

But after this general introduction we should start getting to know our red ants somewhat better, by going into details.

The nest of the red ant

There stands the rounded ant-heap in the silent forest, under the midday sun. . . . A pile of twiglets, countless tiny footsteps along converging paths that are sometimes more than a hundred metres long. Lying on the ground at a prudent distance I listen to them and feel far indeed from the world of men. I daydream of the ants' origins in the forests of the Tertiary era, 80 million years ago. On a planet swarming with life, but where man was not to appear for another eight hundred thousand centuries, the ants already were there, and had been for a long time too, no doubt, very probably doing exactly the same tasks as they do today. They practised agriculture, raised cattle, indulged in the art of war. Why did the flame of intelligence arise in us and not in them? They had a big start over us and had a greater chance of its happening. Or did they, perhaps, a theme taken up a thousand times by science fiction, acquire intelligence as we did, but without our knowing it, a true civilization but one so different from ours that we cannot even see it?

These are daydreams, but perhaps they are the moving force which impels a person to become a myrmecologist. Why does one want to study ants? Unless it is because one is fascinated by them . . . the only animals that seem to behave like men, who do not retreat before him or any other animal, whilst at the same time seeming completely to ignore him: is there not something here to excite a mind interested in the boundless cosmos which surrounds us?

Kicking an ant-heap

Faced with an ant-heap people react differently: the uppermost emotion usually is curiosity mixed with an indefinable horror. . . . But few can resist the desire to give it a kick. What happens then is most strange! The sudden crazy activity, hundreds of ants appearing and falling into the hole, hordes of them running from all over the nest and pouring out from its depths among the twiglets. . . . After a few minutes calm is restored and a persistent and precise repair operation is set in motion. A few hours later no trace of the damage done by your kick can be seen.

This makes a good starting point. Anything can be used to start a train of scientific thought. But a kick is, nevertheless, rather a severe method of commencing an investigation: it has the defect of disorganizing the ant-heap and, in particular, of destroying the outer layer of fine particles and mixing them with the coarser layers of bigger twigs deeper in the nest; moreover, a certain number of ants are destroyed and the widespread shock is too violent a thing to be analysed properly.

Let us then gently push a fist into the nest so as to make a depression two or three centimetres deep: the disturbance is much less; and if we are careful to attack the ants during the heat of the day, when they are in the depths of the nest, we are not likely to get more than a few bites.

What happens then? At first, nothing; but as the depression we have made is hardly visible, we must mark it somehow so as to be able to see if any changes are made in it later on. Here is just what we want, a forked twig lying on the forest floor. We can push this into the soil near the nest in such a way that one of the arms of the fork just touches the bottom of the depression: if it is filled in we shall soon notice it, because the end of the twig in the depression will become covered.

This is just what happens. By the next day, even if the depression was only small and not very obvious, it is always filled up and the heap once more made quite regular. This fact has several implications.

Constant supervision of the heap

First of all the ants must discover the disturbance to the surface and after that they have to repair it. Now, firstly, how have they discovered it? There are not many ants on the heap at midday, and the heap can be big and the depression very slight. We must then admit that the heap is constantly and carefully inspected and repaired if any damage is found. But how can this be proved? There we are in the depths of the forest with no measuring rods or complicated apparatus: nevertheless a number of experiments are possible. For instance, a piece of birch bark rolled up into a cylinder and fixed on to a stick stuck in the ground will make a quite suitable sighting tube and will allow us to note on a piece of paper the exact position of a few twigs seen in the field of this tube. *The position of these twigs is never the same* and sometimes these positions change within a quarter of an hour. Another yet simpler means to demonstrate this continual changing of the twiglets is to make up a few letters on the surface of the heap with the pieces of twig; but you must be careful to do it when the heap is free of ants. Well: you will only be able to distinguish these letters for a few hours! You can do the experiment in yet another way, by making two or three concentric circles with matches on the heap (see plate following page 32). These circles are quickly broken up by the ants. Thus we see that the heap which seems so immovable to us is ceaselessly turned over and rebuilt by the workers.

The ceaseless travelling of the twigs

This Otto has proved by a curious experiment. He coloured the whole surface of a nest blue by pouring a quick-drying paint over it. Once the first excitement caused by this rough treatment was over the workers began to bring new twigs and the blue gradually disappeared. But after a while, perhaps a month or two, the blue twigs reappeared. This proves they circulate, that they get into the deeper layers and then are later brought to the surface once more.

The heap has to be kept smooth and clean

This constant remaking of the heap explains a fact so obvious that people generally do not see it. It is that the heap is clean: it is mostly due to this that an ant-heap can be seen from so far away in the forest. There are no dead leaves on it: either they have been carried away or, if they are too heavy to move, they have been covered with twigs. This can easily be proved by putting bits of paper of different sizes on the heap.

But we can go further than that: as the ants are so anxious that the surface of their heap shall be faultless let us try to tease them with rather more difficult problems. Carrying away bits of paper is too easy. What will happen to twigs pushed upright into the heap rather than just laid on the surface? At first nothing, except some general agitation. But a few hours later or by the next day at the latest, all these twigs will have been removed. They are only fragile stalks like matches: but even twigs of the thickness of a pencil and pushed four or five centimetres into the heap are rooted out within a week and thrown to the bottom of the heap.

And now we must stop and consider for a moment, for this is quite a new phenomenon which could only have been discovered by making the experiment. In nature, twigs are hardly likely to stick themselves into the ant-hill and a curious aspect of ant-behaviour might have remained unknown to us. That's why experiments are important, and one should not confine oneself to observation.

Why and how

All very well, but how did they do it? Is it intentional? We must be careful not unthinkingly to credit ants with any fantastic powers, as our ancestors were wont to do. If in fact ants reason, live and think as we do there is no problem, at least from one point of view; but more detailed studies show that ants are not men and do not act as we do. We must find a simpler solution. That is the famous rule, "Theories should not be multiplied unnecessarily", which the moderns think was postulated for the first time by Mach but which, in reality, goes back to Duns Scotus, the famous theologian of the Middle Ages. He declared *"Entia non sunt praeter necessitatem multiplicanda"*; and in our case there is no need to postulate intelligence since a simple combination of reflexes will suffice to explain the matter; and this, moreover, can be proved by experiment.

A simple hypothesis can be advanced, but I am not sure if it is the true one or the only possible one. It would appear that any twig pushed perpendicularly into the heap releases a digging-out reflex at its base and the thicker the twig and the deeper its penetration, the stronger is the reflex. After a short while the stick tilts over, falls and the ants can carry it away. I cannot prove that this is what in fact happens, but given what we know about ants, as will be shown below, it seems very likely.

Well, this is a long chain of reasoning to be started by a kick! But that's not all. . . . We know now that the ants constantly survey the surface of the nest. Consequently, it is not in the least strange that they noticed the small depression we made. But why and how do they fill it up?

It is not easy to find out. We must do as one always does in such cases in biology; approach the enigma obliquely, turn around the subject instead of attacking it head on. First, one might reflect that the opposite of a depression is a protuberance, which would present another kind of irregularity. Well! Let us put a handful of twiglets somewhere on the heap. The ants will remove it after a while, but without much enthusiasm, so it seems; several days will pass before the last traces of the lump are removed. Consequently it seems that *concave irregularities are more important than convex ones*. But convexities are interesting in this case, because they show that *ants can work not only by adding, but also by taking away*, which is very important, as we shall see below.

The fear of hollows

Up to what point can this fear of hollows go? You have only to place on a heap some empty tins, or better, some empty wooden or cardboard boxes, to see that, for at the end of a few days, they are full to the brim with twigs. You can even make the experiment more complex and interesting by putting on the heap a number of cardboard cylinders of decreasing diameter, making it look something like the tower of Babel! You may have to wait a long time, but, if the weather is fine and warm, the ants will work without cease and at the end of one or two weeks all the cylinders will be full to the top: here the fear of hollows is pushed to the point of a mania.

Well, the cylinders are relatively enormous and easily seen. But when a small depression in the surface is filled one may well enquire what senses enable the ant to notice it. The sense of equilibrium perhaps? Many insects can sense minimal differences in the slope of a surface, 2° or 3°, for instance, with cockroaches, and the ant need be no exception. The use of sight is not very likely, because, like most insects, ants are somewhat short-sighted.

We do not really know why ants dislike hollows in the heap, that is whether there is some biological reason for it deeply rooted in their evolution. But thanks to this powerful force it becomes possible to explain several difficult phenomena. For instance, the shape of the heap: it seems quite natural to us, because we imagine to ourselves a person emptying out a sack of twigs and bits from a certain height, and the resulting pile would quite naturally take on the appearance of a more or less pointed ant-heap. But the heap is not built in this way. Slowly, with difficulty the tireless worker ants jolt along their road, often nearly a hundred metres long, for hours and days on end, dragging the twigs to the nest. And under these conditions there is no special reason why the heap should take the shape of an even, domed cone: there could be a number of peaks separated by valleys. Except that the workers detest valleys! They cannot see one without at once filling it, and this is why the whole takes the shape of a regular dome.

The partition problem

In this same field I spent much time trying to understand an experiment the explanation of which is really fairly simple. On the top of a heap (the top is the "most sensitive part", the area most used by the ants), I placed a cross made of four plates. A few days later the results were unmistakable; there were a lot of twigs in one or two of the quadrants and much less in the others.

What now had to be found out was whether this anomalous behaviour was constant: all that had to be done was to lift off the quadrant and allow the ants to restore the regularity of the heap (which they quickly do), then to put it back in the same position. Would the main heaping-up then take place in the same quadrant or in another? There was no constancy at all. I will not go over the false trails of laborious reasoning I followed before understanding it. You have to be patient and with pencil and paper note one by one the tracks made by several hundred ants carrying twigs and pieces to the nest, with all their stops, turns, retracings of the way and abandonings of the journey: the usual inexhaustible patience of a biologist was put to a severe test. A pretty clear fact emerged from all this confusion: the majority of ants carry their burdens straight before them, to the top of the heap. If then some particular working party brings in a bigger supply on one side than another—which changes according to the day and circumstances—then a rather bigger pile will accumulate on that side of the heap, thanks to the workers whom chance has happened to bring there. *But no irregularity of shape will develop because of the horror the ants have of a hollow.* They will smooth out the lumps and fill in the hollows, so much so that the heap remains regular and rounded: except if you put a wooden cross of partitions on the top, which stops them carrying out their levelling work. Even so they can do it if you bore a few holes in the partitions; communication is re-established between the quadrants and the pile of pieces is everywhere the same height.

Well, this is a long chain of reasoning to come from a kick, and shows how ants by means of a few pretty simple reactions manage to complete complex tasks. But do not for a moment think that all is clear and understood. We have hardly made a step yet towards understanding that other world, a new cosmos, different and shrouded in mist, that of the social insects. We have yet to explain how each individual comes to help at the common task, and it is not easy to do so from the same starting point. We will try another one in a moment, but let us stay a moment longer looking at this ant-heap: there is one fact I left out above, saying I would come back to it. When you place a handful of twigs on the heap, little by little they are removed. This is more important than the filling in of a depression. You might object that at first sight you cannot really see why it is more important as both facts seem to be of equal interest.

Not at all! But to understand why, we must go back much further.

Stigmergy

This word comes from two Greek roots, one meaning incitement and the other work, and together they more or less have the meaning, "The work that stimulates the worker". The idea and the word were put forward by Grassé, when dealing with the immense structures made by termites, to get out of a worrying difficulty as old as the subject of social insects. In what memories are deposited the plans of these vast edifices, so complex yet so constant (remember that some of the mounds of *Bellicositermes natalensis* have a circumference of several *hundreds* of metres). Do the workers themselves, or some of them, retain them? Yet how can we suppose that such a tiny brain, no bigger than that of most non-social insects, and, moreover, insects whose *individual* behaviour is far from intelligent, could conceive any idea, let alone a plan? And even if we admit that they could do this, how do they transmit the information and direct the work? It must be noted that up to the present we have not been able to observe anything like an organized approach to communal construction. Things seem to be quite the reverse and building to take place in the greatest disorder; but what is certain is that the work is done and its proportions are accurate and constant.

Grassé set himself to study and explain the behaviour of termites, put into a vessel with their queen and a handful of rotten wood. The first thing they attempt to do is shelter their queen from light and exposure. Let us leave on one side the first confusion, which is understandable, before building starts. After a while calm reigns once more and the workers start to put little pellets of building material here and there. These grains are stuck anywhere, just by chance, on the floor, and do not seem to interest the workers carrying other building pellets.

But it may happen that a worker puts its pellet not at the side of one already in position *but on top* of it. Statistically, there is a reasonable chance of this happening. . . . The workers' behaviour then slowly changes. They tend to take more interest in this embryo building. There is thus a greater chance that other workers will put their pellets on top once more and start a column growing upwards. This process can be repeated several times and we get a forest of small columns rising from the floor of the vessel, and it may well happen that a number of them are fairly close together. In this case we find not the indefinite prolongation of the column (at some point there seems to be a brake on the stigmergic process when the stimulation has gone beyond a certain stage) but a tendency to place the pellets laterally at the top, that is to make an arc which will join the two columns. So much so that finally the whole base of the vessel, and first of all the queen, is covered with continuous vaulting.

Nobody "intended" anything at all, nor did anyone "have a good idea". A small number of automatic reactions combined together have simply produced an almost miraculous result.

The remarkably ingenious theory of stigmergy is not here

just a hypothesis: it is a fact. It is more than probably so *at the start* of building. But there is one thing! The theory of stigmergy only serves to explain the building of a sponge-like structure. Is the termite nest like this? In places, yes, but many parts are very different. Without going into details, because this book is about ants, not termites, we may point out that the external shell of the termite mound is thick and compact, penetrated only by narrow exit and entrance passages; the queen's cell is oval, very hard and isolated among the spongy mass of fungus gardens; there are also the famous conical pillars of the base, discovered by Grassé (these pillars support nothing at all, since their lower points do not touch the soil; they are better compared to stalactites, but ones made by the workers). We should also mention another instance noted by Grassé, about part of a nest he saw in Africa made of a special kind of clay which could only be got from a layer fairly deep in the soil beneath the nest: consequently the workers had to go and fetch it, passing on the way innumerable sites that did not stimulate them to any building activity, then come back, again passing the sites, which again stayed quite neutral for them. One must thus conclude: (a) that they know the layout of the nest; (b) that when they are seeking some special kind of material, or are returning with it, their building fever is inhibited in most of the sites, except one, and that, momentarily, the work does not stimulate the worker.

This complicates the theory of stigmergy, without forcing one to reject it; for instance, as regards the nest's architecture being different according to zone, one could postulate that not all teams had the same reaction. Possibly, but there is something else: the work, or rather the building material itself, can stimulate the worker not to build, but to destroy.

Darchen, first of all, found this in bees. Everyone knows that these insects willingly use an artificial wax foundation to make a honeycomb; but looked at a little closer, they are not just content to build on it. It is known that when they spontaneously build a comb on their own, having no artificial foundation, the first row of the comb always has the shape of sectional drops of water. Darchen, then, offered the bees a bit of foundation wax marked in squares or rectangles; then by very close observation, because it all happens very quickly, he found that *the workers cut away from the square or rectangle all surplus wax*, so as to leave a piece more or less the shape of a drop of water! We may thus conclude that the building material no longer stimulates the workers to build, but to destroy, at least partially: and how can we avoid the idea of a plan?

Let us go back to the ants: this is why I wrote above that destruction has not the same significance as construction. The ants also take away from the mound all surplus material, such as a handful of twigs put on its surface, although the material is the same as that used for nest building. One can even make a more drastic test; still using only the nest's material, I interested myself in changing its shape, for instance, by narrowing the base whilst increasing the height: the mound became like a pointed hat. The ants, quite understandably, hated this tomfoolery and energetically bit and sprayed me with formic acid; then, after several days (the reader must be beginning to know what to expect), the ants restored the heap to its original shape, by taking away "all that was superfluous".

I can quite readily imagine the objection: when you alter the shape of the heap you are not just restricting yourself to a simple moving around of nest material; you have upset the whole structure of the nest, the distribution of fine material on the surface and coarse at the centre. This is true and is corroborated by another experiment. I remember being particularly obnoxious one day and removing all the nest material until I exposed the rotten tree stump on which such nests are usually built. Then, possibly overcome by remorse, I decided to help the work of reconstruction by myself placing a handful of twigs on the old stump. But the ants, who no doubt bore me a grudge, took them all away and only later did they start to rebuild the nest to their own satisfaction.

Consequently, *they knew how to recognize their own work and how to distinguish it from that of any outsider.* Once more we have added something to the theory of stigmergy: it is no longer the material, the bricks, that inspire the worker, *it is the wall itself,* perhaps because of what it lacks, perhaps because it has too much. We must admit, however, that this brings us no nearer to the concept of a plan, even in a diluted form. I cannot altogether admit that the conclusion is inevitable; I think it more likely that the equipment of reflexes of the builders is more varied than we thought at the start. I believe it has been shown fairly convincingly that the material stimulates the worker, and that the mound in course of construction draws more and more workers in to forward the task; but *once the heap is built other reactions dominate*: a hatred of hollows and humps, and an awareness of a construction in several layers which must be maintained. And such an awareness must exist for, if it did not, the constant supply of twigs and pieces brought in would build up a very irregular surface, which is not the case. No doubt our list of the building reactions of ants is still very incomplete!

Heating among the ants

The first fact myrmecologists note on examining nests of the red ant is that *the centre is warm*, as is the middle of a beehive. On the other hand, from September, even though the ants are still very active, the interior of the mound is cold, more or less at the external temperature. Consequently, we must suppose that the nest can produce its own heat; at least one might think so at first sight. But an examination of the heat question is going to lead us step by step to study the ants' annual cycle, metabolism and many other matters.

Let us start at the winter resting stage, when the ants are deep in the nest, sometimes more than a metre below the surface of the soil. What becomes of them at this period?

Red ants in winter

In the galleries dense masses of ants are found whose temperature scarcely rises above 0°C if it is freezing outside: this is the first difference, and an important one, from bees, who manage, even in very cold weather, to keep the hive temperature well above that reigning outside. The descent to the bottom galleries coincides with the fall in temperature of the mound, which starts, as we have seen, from September and rapidly increases. At the same time there appear among the workers individuals whose abdomen is enormously swollen and filled with protein and fat stores. They constitute nearly a third of the population and no doubt are derived from the young workers hatched that season. An analogous behaviour pattern is found in bees, but there it includes the whole population: the winter bee, living more than six months, is very different from the

summer bee, short-lived and with none of the abundant food
reserves found in the winter bee's body. It is plain that the
ants' winter stores are to some extent equivalent to the pollen
and honey the bees put into the comb. Though their ultimate
use is very different, as we shall see.

Winter movements

In the ants' nest there is a special and strange kind of activity:
the moving of these ants stuffed with food, or "reservoir-ants"
(the *Speichertiere* of the Germans), by other ants, of the external
service, into the depths of the nest. In truth these movements
of one kind of ant by others take place all the year round, but
there are two peaks, one in the spring and the other in the
autumn; in spring they are moved either to the interior or to the
exterior of the nest, to daughter colonies, for instance. The
worker thus moved folds up its antennae and legs as if dead. In
spring at least we do not know what the release mechanism for
this strange behaviour is; in any case the number of workers
moved can reach 40,000 for a single nest. According to Zahn
this movement could be in proportion to the summer needs of
the colony, but he thinks this still needs to be proved.

In winter the situation is more obvious, but nevertheless
we may note that these swollen workers are not helpless and no
doubt could move of their own accord. The moving-men (or
rather moving-women) belong to a well-defined class of worker,
the exterior workers (the *Aussendiensttiere* of the Germans);
they have no food reserves and enjoy the peculiarity of remaining a little active even at very low temperatures of the order of
one or two degrees (C). This is sufficiently strange, for though
many insects can survive very low temperatures without
difficulty, they soon lose all power of movement. As far as I
know there is only one other insect which can live almost at
freezing point; it is the famous *Grylloblatta*, living at the snow-
line of the eternal snow, which very quickly dies at the least
rise in temperature. In addition to the ant-reservoirs the interior
service workers are also carried into the basement. The result
is that the winter rest has a fairly complicated population
make-up. We can find: (1) females; (2) reservoir-ants with
swollen abdomens, important as storage vessels, ovaries well-
developed, poison vesicles empty, reduced transpiration and gas

exchange; (3) exterior service workers, poor food reserves, ovaries totally absorbed, poison vesicles full, considerable transpiration and gas exchange; (4) interior service workers, intermediate in body development and metabolism. Moreover the distribution of this population in the nest is not equal: the exterior service ants are found in great numbers in the outer layers of the nest where they are able to support without harm temperatures of – 30°C and a great number of successive frosts and thaws.

Spring

When the rays of the spring sun start to warm the nest the ants may be seen moving little by little into the mound's upper layers, which evidently are the warmest. The exterior service workers again take up their task of moving their colleagues, but this time it is upwards. At this time, March-April, one can see great masses of immobile ants on the nest: the Germans say they are in a state of passive sunning (*passive Sonnung*). With *polyctena* even the queens take part in the process. But it is not easy to understand the reason for this: it only happens on "cold" mounds, which at times are a little colder than the surrounding atmosphere, that is to say they have not yet adjusted to the reigning temperature and have not been much warmed by the early sun. By contrast, the phenomenon is not seen on warm mounds, which have reached the ambient temperature; here there is no passive sunning, but only busy workers running over the surface.

What has happened? To understand it we must have recourse to the great physiological changes taking place in the workers, starting with their "thermo-preferendum".

The thermo-preferendum of ants

This barbarous word simply means the *preferred temperature* of ants. The preferendum is one of the most widely used things in animal psychology: we have not only thermo- but also photo- and hygro- preferenda. To measure preferences the creatures are presented with what is called a temperature, light or humidity gradient: this is a place where the heat, light or humidity increases regularly as the test creature moves. To measure the thermo-preferendum Hester devised a simple apparatus to which he gave the strange name of a "thermal organ"; it was a long metal plate heated at one end and cooled at the other. The test creatures were put on it and came to rest at the spot they "preferred". In fact, it is unlikely that there is any preference; simply that, and it can be proved by experiments which I do not have time to describe, the amount of an animal's movement is much stimulated by temperatures below and above the thermo-preferendum; at the preferred zone, on the contrary, their activity is much less, so much so that statistically one has a greater chance of finding them there: the thermo-preferendum works as an "activity-trap".

Now it is well known to all myrmecologists that in spring the ants will collect their building materials, eggs, larvae and pupae under a lamp bringing the temperature of the substratum to some 30°C. But the matter was further studied by Kneitz and Zahn and it is not simple; for instance, the thermo-preferendum depends on humidity (a factor to which many myrmecologists give more importance than to heat), on food and on season of the year.

Above all the thermo-preferendum is *far from being the same for all individuals*. In summer, for instance, bees on outside duties show practically no preference; on the other hand those working inside have one, but it is not very definite, with a tendency to group at the cold end. We should note that in bees too the thermo-preferendum is not well defined.

It thus seems that in the ant-heap isolated individuals are scarcely influenced by heat. *But everything changes once the brood is introduced.* The reaction then is emphatic and clear: the workers mass at the point showing 25°–30°C and the queens prefer 21°–22°C.

In autumn everything becomes as before; queens and workers no longer gather with the brood around the warm spots, *but avoid them*: the time for hibernation is near; the workers look for a temperature between 4° and 5°C, which is about that of the hibernation chamber. Nevertheless, some workers are still strongly thermotaxic: these are the "thermic messengers", who, at the end of winter, go outside and warm themselves in the first rays of the sun, then descend again to the hibernation chamber, carrying a few calories with them. As the strength of the sun increases they grow more wily and will *even carry workers out of doors.* They play a more important part in the "passive insolation" process than the slow warming due to the rays of that globe. Finally, the majority of the workers, but not all of them, come outside and the activity gradually becomes that of spring.

Zahn has been able to show that the workers are able to pass on thermotaxic news: he used an artificial nest more than a metre long, at one end of which was a low heat source. After a while all the ants previously scattered throughout the nest gathered beneath an opaque felt cover at the warm point; this means that those ants nearest the warm spot had somehow informed the others.

Migration of the sexuals and the start of breeding

Another migration occurs at the end of the sunning period: the queens get into the deeper and cooler part of the nest. And, at the same time, the workers undertake the raising of the sexuals, as if the withdrawal of the queens had left the brood as "physiological orphans". The situation is comparable to that of a beehive when the queen is removed and the workers take over the care of the royal cells. At this time too the food reserves of the swollen workers we have already mentioned are brought into use. They are gradually introduced into the system and the reservoirs are "empty" from May until June; but the young workers have already hatched and their fat and protein content continues to rise until the start of winter. The heat regulatory mechanism also starts, which we will now consider.

Heat regulation

Heat regulation is not achieved at once or completely in each nest. Optimum nests can be found, with a high mound, well populated; non-optimum nests, inhabited by young colonies; or, on the other hand, old nests split into several parts, or pulled down at some time by man or animals. Here the mound temperature does not remain at the optimum for very long, and it reaches it only when the sun heats the mound. Even when the nests are well populated and apparently healthy, there are some which have a good exposure and are heated for a considerable time, and regularly, by the sun. The mound can even reach a temperature of 70°C, but it drops rapidly as soon as you penetrate the interior. There are also nests, shaded and with poor exposure and only getting the sun for short periods, where the temperature of the mound varies a great deal. Finally, there are nests situated in the shade where the temperature of the mound is that of the ambient air. But if the colony is healthy the temperature, not of the whole mound but of its central part, changes very little.

The above observations were made by the Belgian myrmecologist Raignier who gave great importance to the sun and but little to the workers. Not everyone, however, is in agreement on the relative importance of the sun and metabolic processes to the temperature of the mound; Heymann, for instance, noted a

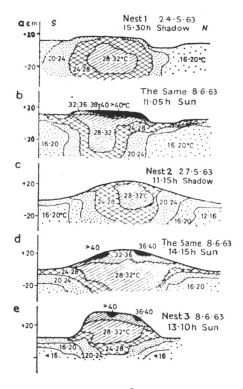

FIG. 8

Isotherms in nests of *Formica polyctena* at different times and places; no wind (after Kneitz).

dozen cases where the temperatures of the mounds at depths
of 10 and 15 cm. *rose* when the solar radiation *was dropping*;
and in other cases the temperature of the nest *fell* whilst the
solar radiation greatly *rose*. We concluded by advancing the
proposition that external factors, such as the sun's rays, have a
certain influence but not a determinate one. We should also
note that temperature regulation is not established, nor is it
maintained in *all parts of the mound at the same time*. Isotherms
noted daily show that at first the zone where the ants control
the temperature is small and circular and this zone increases
in size as the temperature and amount of sunlight increase.
It can well get smaller in cold weather (Fig. 8).

Finally, often on cold nights an enigmatic *increase* of temperature is noted, since on warm nights a considerable proportion
of the ants go foraging. Steiner supposed that on cold nights all
the ants stay in the nest, so that the temperature rises and the
fall in this figure in the early morning corresponds with the
departure of the foragers. Raignier, on the other hand, thinks
the mist forming around nests on cold nights makes a screen
preventing loss of heat from the mound by radiation.

It will be seen that there is more than one problem still
to be solved on the subject of thermal conditions in the nest,
and it is not yet finished.

Humidity in the nest

Humidity is very high, nearly at saturation point, and seems
strangely dissociated from temperature changes, no doubt
because the workers bring in water. At night the humidity
drops, often due to the nest's heat, which induces evaporation,
when the addition of water, by workers, is nil or reduced;
later we will note the importance of a high humidity in temperature control.

What is the heat source?

At the present point of our discussion we may well enquire
what is the source of heat production; is it the sun? Or do
ants, like bees, produce their own heat? Up to the present we
have not found any compelling argument for one view or the
other.

First of all, we must do justice to some old theories more of historic interest than any other. Such as Forel's theory of the mounds, which he considered to be accumulators of solar heat. But one simple observation casts doubt on this, for the dome is warm even during prolonged dull weather; moreover, when a house is covered with thatch it is not for the purpose of capturing heat from the sun but for protection from it and to stop internal heat from escaping. . . . Zahn's thermic transfer theory (1948) is much more recent and rests on what *seem* to be self-evident arguments. This author saw the passive sunning of the ants in early spring as essential; according to his calculations the warming of the ants' bodies was sufficient when they went back into the nest to warm it to any desired extent. But Kneitz objected to this on the grounds that there is perfect temperature control on cloudy days and at night, and in nests in the shade. Moreover, if you fumigate the nest and kill the ants you find that it cannot maintain the heat characteristics of a living colony. This last fact seems to be decisive, although the transfer of heat directly from outside must not be neglected.

The only theory now left to us is *a heat source in the metabolism of the ants*. But the situation is very different from that of bees. With these last, for instance, heat is produced in the highly developed muscles of the thorax. But with ants, particularly with the completely wingless workers, there is nothing like this and their metabolism can only raise the temperature of the nest when the outside temperature is not too low, that is to say, in our latitudes, between May and September, and in a protected position at that, resistant to heat loss, like the mound. Kneitz found an *unexpected limiting factor—humidity*. Here I am forced to make some simple calculations, but I think they are worth while because the result is, I repeat, astonishing enough.

Let us take a nest of a million workers with a volume of 100 litres. If we take the respiratory quotient as 1, little different from actual measurements, then glucose metabolises at 5·05 calories per litre of oxygen. Now at 20°C the workers use 700 mm^3 of oxygen per gram per hour; and a million ants (about 9,500 gms.) 6·65, that is 33·58 calories. A special characteristic of our red ants and of many other species of European ants here comes into play. *They transpire very greatly and seem incapable of retaining water in their system*, when the atmospheric humidity is under the saturation point. At 95 per cent humidity and 21 °C they get rid of 0·05 mg. of water per worker per hour; this is 50 g. for a million ants, which needs 29·3 calories to evaporate it. Consequently, we have 33·58 —29·3—4·28 calories per hour available for heating purposes. This is enough suitably to heat 100 litres of air. But only *on condition that the humidity is kept very high.* At 75 per cent humidity and 21 °C the workers would lose 0·13 mg. of water per worker per hour, that is nearly three times more than at 95 per cent (130 g. per hour and 76·18 calories for a million ants) and then the nest can but cool rapidly. On the other hand, at higher temperatures equilibrium can be maintained, always provided the humidity is within suitable limits. For instance, at 25 °C the oxygen consumption is 950 mm³ per gram of ants per hour, that is 45·38 calories for a million, and at 30 °C it is 1,200 mm³ and 57·57 calories per million.

Thus the main purpose of the mound is not to accumulate heat from the sun but *to stop the loss of internal heat and to maintain a very humid microclimate inside the nest,* essential for heat control there. Moreover, when the weather is very hot and dry, the mound may well not suffice for its task, because it is porous; if this happens in early spring the colony cannot produce its sexuals. No doubt this is the reason that in hot and dry regions nests are nearly beneath the ground or even completely so, with no mounds of twigs above them: to maintain the humidity they have to go down into the earth, the dome is not enough. (*Formica nigricans* in the Causses region, for instance). I must add that in the case of overheating or of excessive humidity a certain amount of extra ventilation can be obtained by the ants enlarging the nest openings.

How do they feed themselves?

Transport of prey by ants. For a long time myrmecologists have been fascinated by the tiny pattering of innumerable ants along their paths: in summer it never stops day or night—the workers, one carrying a dead prey, another a twig and yet others some honeydew in an extended abdomen. But one phenomenon has given rise to very bitter controversies: this is the *make-up of the collective efforts a number of ants engage in to move a very heavy prey.* The unskilled observer might think they are all helping each other as would a number of men, and that is what was put forward in former days when anthropomorphism flourished unrestrained in biology.

A study in greater depth, however, at once shows certain difficulties in accepting a theory of mutual help as practised by man. For instance, how many times do we not see ants pulling at something in different directions? Observing these apparent absurdities, the French biologist Rabaud, who had his hour of glory for his very intransigent views, maintained that each ant, or any other social insect come to that, really worked only for itself, without worrying in the least as to what its companions did. An absurd view, it is true, but one which had the merit of starting further studies. As usual the truth was not as simple as was supposed by either of the parties . . . (see p. 197).

I thought at that time that the controversy could go on for ever if one did not take care to define the terms being used in the dispute. For instance, the words "mutual help" evoke the idea, for us, of conscious deliberation, and obviously this was not true in the case of ants. But in any case an objective definition should be possible. If mutual help in fact takes place then two should carry the same prey better and more quickly than a single ant.

I soon established that this was the case. Sudd, later, took up the same problem, but with a better technique than mine. For instance, he attached a glass fibre to a prey at one end, fixing the other to a firm point, thus making a kind of dynamometer or simple torsion balance. The deviation of the glass fibre, measured by means of paper ruled in millimetres, allows the force used by ants in dragging the prey to be measured. The device confirmed that transport is easier when several ants attempt the task than when only one does. But one must distinguish between different kinds of ants: and there are other aspects too, which are more difficult to explain.

Let us pass over the first stage, where the ant licks and bites the burden to be transported; and also omit the case of small objects, carried by one ant with the head raised high, and not pulled or pushed. Here it is we find that well-known attribute of strength in ants: the large *Myrmica rubra* workers, for instance, weigh 2 mg. and can carry 7 or 9 mg.; the smaller workers, scarcely more than 1·5 mg. in weight, can at times do the same. But it is said that this strength is more apparent than real, and if you consider the relationship between the weight lifted and the volume and section of the muscles used, the ant is not stronger than man. However well-based these calculations may be, the muscular strength of ants is nevertheless remarkable.

Lone ants frequently change their grip, let their burdens fall, clean their limbs, then take up the object again, catching hold of it at another point, particularly if obstacles have hindered their progress. The same thing happens if two ants, and not one, take on the same prey: Sudd's ingenious glass-fibre dynamometer, however, shows that on average the tractive force is at least twice as much when two workers undertake the task; but only on average. For in 30 per cent of the cases, the dynamometer shows that only one of the workers is really doing anything, the other is content to lick and bite the prey. But in 70 per cent of the cases they effectively collaborate, though with greater or lesser efficiency, as Sudd remarks. The variations in efficiency measured turn on the question of direction of the two tractive forces in respect to each other; the direction of their efforts may vary considerably without going quite so far as each to attempt opposite directions. It is strange to note that the greatest tractive force is only reached after several minutes of combined effort as if the workers progressively stimulated each other. This is not just an airy theory, for, according to Sudd, lone ants carrying a load show, on the contrary, less and less energy as time passes. And, moreover, one often sees stoppages, lost time, or two ants each dragging in opposite directions until the stronger one dominates; the second ant then follows the first after hanging on to the prey as long as it can. This lost time is particularly often seen during the first stages of transportation, and much less during the second. For during this second stage the ants show a marked tendency to drag their prizes in the direction of the nest.

Is this all there is to be said about the ants' transport powers? Not at all, for Meyer, working on the red wood ant (whereas Sudd was interested in another species), found that far from the nest loads are often abandoned, lost or carried in the wrong direction. Close to the nest (some ten metres) is where the workers' tracks become less devious. Here the ants are most likely guided by the sight of their numerous companions travelling along the narrow paths, which, to some extent, help to keep each individual ant "on the rails". As Meyer says, this "traffic signal" is very strong near the nest and points always in the right direction; away from the nest it is lost in the silence of the forest and the porter ants do not seem to have as good a sense of direction as their unloaded fellow-workers.

Time of foraging

Naturally if we are going to use red ants for pest control in the forest (see p. 68), we need to know exactly what they bring back to the nest. The first idea that occurs to one is a little naïve: it is to sit as near as one can to the foragers and to take from them everything they are carrying; it is yet easier to take a sample of ants at random, anaesthetize them and pick out the various pieces of prey they have. But both these methods produce violent upsets in the colony and can only be used, especially the second one, at somewhat too long intervals of time. Turning over these difficulties in my mind I hit on the idea of an automatic trapping device: the inspiration was a simple little piece of apparatus used in bee-keeping, the *pollen trap*.

Bees bring back to their hive two large balls of pollen attached to their hind legs, which they have collected grain by grain from the stamens of numerous flowers. If you put a plate pierced with 4 mm. diameter holes in front of the hive entrance, these holes will just let the worker bees enter but the large lumps of pollen will be scraped off and fall, after a while, into a drawer placed beneath. To speak the truth things are not quite as simple as that and nothing is more interesting than to watch the efforts of the bees struggling to get their prized pollen masses into the hive, sliding one foot after the other through the holes; the strangest thing is that they succeed and slowly the yield of the pollen trap drops. Nevertheless in this way a strong hive can give you 200 grams of pollen a day or more. . . .

Consequently, I said to myself: one should be able to play this sort of trick on ants by making them pass through some suitable hole which will hold back their loads. After a few attempts I managed to perfect the apparatus. Fig. 9 explains the principle. The workers are separated into two columns, those entering and those leaving; only the first interests us at the moment. The nest is surrounded with cardboard soaked in fuel oil which

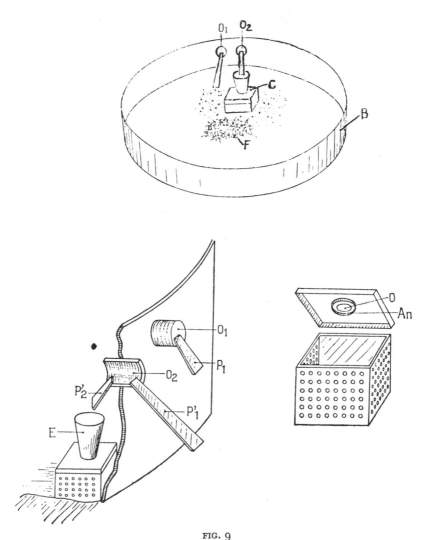

FIG. 9

Above: apparatus used to measure foraging activity. B. Circular barrier soaked in fuel oil; F. Nest; O_1, O_2, exit and entry points; C. The "game catcher", beneath the entry point.

Below, left: E, metal tin into which the ants fall after having travelled along the path P'_1, P'_2, the re-entry path touching the ground outside; O_2. Entrance; O_1. Exit with shortened track P_1, allowing the ants to leave by falling from the end of this track, but preventing them from entering this way;

Below, right: Enlargement of the collecting vessel with cover raised to show entrance O, surrounded by a cardboard barrier, An, soaked in fuel oil. The cover itself is treated with fuel oil to prevent the ants leaving the vessel this way (after Chauvin).

forces the workers to use only one entrance: this single gangway touches the ground outside but not inside the enclosure. The workers get over this gap very easily by letting themselves drop: an action often carried out by red ants. But they cannot leave by the same route because they are no "jumping jacks". However, they have another passage-way inversely fixed, which allows them to get out by falling into the outside world. The whole thing is very easily set up and nothing is stranger than to see the double file of foragers moving in opposite directions at three centimetres' distance.

Automatic collection of loads

I next placed under the entry staging, inside the enclosure, a square box, the cover having an entrance and the sides being pierced with holes just of a size to allow the passage of an ant. The workers carrying burdens have to let themselves fall into the receptacle, scramble around in the box and leave it by the lateral holes, leaving their loads inside—that's all, if the god of myrmecologists is in a good humour . . . and he is! I installed this box at nine in the morning in the month of July and to-wards 8 p.m. I found a motley crowd of insects or bits of insects of all colours, sizes and shapes: and I was very pleased. Especially since in the explanation above I have glossed over some of the preliminary difficulties. For instance, in this type of trap the least detail is of significance and I learnt, to my cost, that red ants detest metal and plastics. If the box is made of one of these substances the ants fall into it readily, and fill it to the top, but they refuse to leave it; they all pile up and die from crowding and the formic acid which the excitement causes them to discharge.

Another strange thing noted is that theoretically only big burdens would be held by the holes: there is no reason why small pieces, no bigger than an ant's head, could not go through them. However, they do not and a mass of minute insects, intact what's more, is found there—tiny chalcids, for instance, scarcely visible to the naked eye. No doubt the confusion inflicted on the ants is so great that they let go of all their burdens at once.

Thus, the first difficulty was overcome. We shall see later on that these results do not agree too well with those obtained by the Germans, using hand collection. It is not surprising; for this often happens when one thinks one has perfected a new technique.

Forest ecology

But there is another problem, not an easy one at that. The scientist is not satisfied with merely noting the "green islands" around the ants' nests in the forest after an attack of pests. He wants to know in detail just what the ants have done. Only by knowing this can he perfect a technique for field use. The first thing to know is *what kinds of prey the ants take and in what quantities*; or, more exactly, given a known population of a forest, what part of it falls to the red ants?

The difficulty is in the expression "given a known population"; because up to the present we do not have a reliable means of establishing this. That, moreover, is always the difficulty with *ecology*, the science studying animal populations in the wild. But, you may object, why worry about that? When there is an abnormal increase in a pest species, will not counting the pests the ants bring back to the nest suffice? No, because, if there is no quantitative measurement of the infestation, how can we measure the *relative efficiency* of the ants' action? It is statistically certain that they cannot fail to bring back an appreciable number of any insect whose numbers greatly increase in their foraging zone; but this is not definite proof. On the other hand, these insects are themselves parasitized by others which constitute an effective brake on the pest's increase. Now it has been noted many times (and I have seen it myself) that ants bring these "useful" insects to their nests as well, *thus working against their beneficial effect for man*: but to what extent? Here again we need to know from the first what the population of useful insects is; a second problem no less prickly than the first. But, our stubborn adversary may insist, these are just entomologists' quibbles. Is it not enough to note the existence of "green islands" and then to establish enough ants' nests to make the whole forest one verdant isle? Unfortunately, no, for, as Gösswald saw at once, "there are ants' nests and ants' nests and woods and woods"; *very big differences* in effectiveness can be noted, which brings us back to the problem of the ecological explanation: once again *we must understand before acting*.

Selective foraging

Now, methodical observations, made by a host of observers over years, of the victims captured by ants show *that these do not take at random all that comes to their jaws*: they have preferences and the hunt is highly selective. Two points at least seem contradictory, though both quite true: firstly, ants that hunt by sight are much drawn by prey in movement and throw themselves on to such moving insects in great numbers; Wellenstein has noticed, for example, that ants attack the pupae of butterflies and moths because, when touched, they usually "lash out" with the end of the abdomen. On the other hand parasitized pupae, which are half paralysed, do not respond to touch, and are left alone. Ants catch large numbers of flies, among them many syrphids, voracious destroyers of aphids: as we shall see later, ants will go to any lengths to protect their cherished aphids. Now the flies are very lively and have better sight than ants; admitting that ants are attracted by the flies' lively movements, how the devil can they catch them in such large numbers? Finally, and this is what seems contradictory, the ants go *hunting at night*. I was not the first to notice the red ants' nocturnal activity, for many authors have remarked on it before me; but I think I am the first, thanks to the automatic trap, to publish lists of what they bring back. Now on the darkest nights the percentages of species captured does not vary much from that of their daytime victims, though the total of prey captured drops. *At night* the ants cannot use sight; consequently it is clear that they do not have just one method of hunting but several.

Special tastes

As regards special tastes, here are two examples. My wood in the Ile de France is full of wood grasshoppers (*Nemobius sylvestris*), whose faint, sweet song can only be heard by young ears. Alas, I can no longer hear it! But for me it was once the very music of the forest. Now *Nemobius* ranges everywhere right through the foraging zone of the workers, and yet you may only find one or two in the automatic trap: an insignificant proportion when one considers the size of the actual population. A second example: ladybirds. Red ants, according to all the

German authors, leave them alone, on the one hand because of their rounded shape, making it difficult for the ants to get a grip on them, and on the other because of certain repellent secretions. But I cannot confirm this: in fact, ladybirds are never absent from my traps. Finally, there is a quarry which the red ants like above all others—other species of ants. At the time of swarming they have a good blow-out on the fat females of *Lasius*, bringing them back by hundreds every day.

All this makes a complicated picture. Now let us see what the Germans say about pest attack in the forest.

Red ants and forest pests

In the literature one used to find, and one still does, authors who do not care to admit that red ants play any useful part in controlling pests. This is due to the fact that the necessary distinction between *Formica rufa* and *F. polyctena* has not been made. The former has large nests with but small populations (100,000 workers at the most), and the latter has nests much less voluminous, but far more highly populated (one or two million workers). You must be on your guard when relating number of nests to the state of the vegetation, not to forget that there are nests and nests! Moreover, the *polyctena* nest, with its thousands of queens who can reproduce themselves, is immortal, at least in theory.

Along the tracks go vast numbers of workers, several hundreds of thousands a day along each path: not all go to the end and some even turn back for reasons we do not really know. All the same when a large-sized capture has been made a considerable number of ants assemble to cut it up. We do not know how workers are recruited in the ants' world, as up to the present no one has shown that there is any mechanism comparable to the bees' dance, by means of which the scout bee assembles her companions and tells them the direction and distance of food supplies. However, we have the "kinopsis" mechanism of Stäger, which seems well founded: an excited ant, or one moving quickly, soon attracts the attention of other ants and induces them to approach it. Now a worker, having found a large victim, runs here and there in a characteristic jerky manner. But if this mechanism works during daylight, what happens at night? We still have the problem of night hunting and the senses that control it; I have already alluded to the fact that we know nothing about it.

The fluctuations of the hunting urge

In addition, not all the workers seen running along the tracks are in the mood for foraging; far from it. The biologist Lange put some Tenthredinid larvae, *Diprion pini*, a pest of pine-trees, near an ant track, and found that out of 81 workers who passed by 23 seized the prey and carried it off, 16 attacked it vigorously without overcoming its resistance, 28 seized it to let go immediately, and 14 took no notice. If the creature is dead the ants are even less interested. Out of 240 workers offered a dead *Tenebrio* there were only 76 who took it against 167 acceptances if the creature was alive. Another very good example of the influence of a prey's being alive and moving can be seen in the case of *Lygaeonematus abietina*, a hymenopterous pest of pine-trees, the larva of which scarcely moves except just before spinning its cocoon: it is at this moment that the ants attack it. The cocoons, on the other hand, are just as numerous all round the nest, and up to some tens of metres' distance, but the ants do not touch them.

Much also depends on the *amount of stimulation* to which the ants have been subjected, which increases the nearer they are to the nest (as no doubt does the desire to forage): the distance at which a not very mobile object attracts ants decreases as one draws near to the nest. We do not know if the number of ants foraging depends on the spoils they find; or if, as occurs with many animals, it is merely a question of automatic exploring, with the prey thrown in as a bonus. Up to the present nothing has been found comparable to the expeditions and periodicity of the army ants studied by Schneirla (see page 71). All the same, Ayre, using an American ant, *Formica subnitens*, related to our red ants, found a decrease of foraging activity at the time of the pupation of the brood. We must also add that aphids, moving from their usual sites, do not escape the powerful mandibles of the ants.

Weather influences ants

Hunting starts towards 8° or 10°C, but its intensity and yield increase with temperature. Several authors have found it is greatest around 10 a.m. and this is also the maximum period of ant movement, which we must now discuss.

The general activity of ants

As I have already said this activity has surprised a good number of authors, fascinated by the disciplined columns of ants plunging into the depths of the woods and coming back loaded with food. But when one tries to calculate the numbers difficulties arise, because the ants are too numerous. One gets lost! With whatever system used, at the slightest inattention the number is lost and the exercise must be started again. Some automatic method is needed.

But, as always, deep in the woods I have scarcely anything available, except in a small workshop that I have at my country cottage. Twenty-five years of research without financial support have taught me more than one useful lesson, as I said above: for instance, that there is always some way of overcoming *any experimental difficulty* provided one is prepared to use the little grey cells, as Hercule Poirot used to say. You must know that before the electronic era we used to pass round tips in our laboratories on how to make a whole series of gadgets, some very sensitive and of remarkable precision, which cost nothing, the main thing then for us. For instance, do you know how to make an ultra-sensitive balance, frictionless, with only a minimum of material—a bit of copper wire, two needles, a piece of tin and a small iron plate? It is very easy: as Fig. 10 shows, you solder the needles on to the bit of tin and you rest the points in two cups made by a steel punch in the iron plate. Then, fixing on the copper wire (suitably curved in order to lower the centre of gravity), you have your balance: its only contact with its framework is the two needles. With a similar bit of apparatus, and using the reflection of a ray of light from a mirror fixed to the balance, I have had no difficulty in recording the movements of a flea. Here, in the case of ants, the problem is less complex, for they are relatively heavy for such a balance as ours. But we still need a long recording stylus; a dry straw will do. As to the recording cylinder, a Paris firm has, for nearly eighty years, made such cylinders, worked by solid, reliable clockwork, which our fathers used, as also our grandfathers and which, no doubt, our great grandchildren will use too!

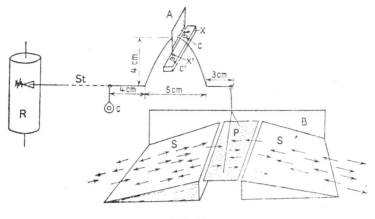

FIG. 10

Apparatus used to measure ant movements. The arrows represent
ants. S. Gentle slopes leading the workers to P, balance pan; R.
Recording cylinder; S. Recording stylus with counterweight C;
A. The frictionless fulcrum; X, X'. Two needles reposing in C, C', two
cups (after Chauvin).

It is very suitable for use in the open air; and, as the great
Marey, master of the graphic method, demonstrated, any sort of
paper suitably blackened with lampblack and folded round the
cylinder can be used to record the most delicate movements.

Obviously we have not solved the whole problem with this
little toy: we still have to persuade the ants to cross the pan of
the balance. Now they can be very pig-headed (often!): I
suspended a platform made from very thin aluminium or
plastic from one arm of the balances; they refused to cross it. I
had to substitute a leaf of fine cardboard (one of the outer
layers of a piece of corrugated cardboard) and then to spot it
generously with Indian ink to lessen the contrast between the
whiteness of the board and the brown soil of the forest. The
shape of the platform also gave trouble: I thought it advisable
at first to make it long and narrow; like that, I reasoned, a
considerable number of ants will be on it at the same time and
there will be a better chance of moving the pointer of the
balance. But I could not persuade the ants to use it: they
dropped to the ground, seeking outside passages and always
finding them, in spite of my dousing the repellent fuel oil all

over the place. On the other hand a short wide platform was accepted without difficulty.

I believe the reason for this is simple: the flow of ants is always dense, but not to the point where meetings and collisions between workers are frequent. On a long narrow gangway, collisions often occur, whilst the wide short one is like a natural track, or at least very close to it. I go into all these details just to show how to make an experiment in practice and how nothing is ever simple. This, moreover, is a reason why one should make experiments, because something unexpected always comes up, which observation alone would not have revealed.

All these difficulties were overcome in the end and I obtained some good graphs, as Fig. 10A shows; you can imagine my satisfaction. It turned out that the balance was more sensi-

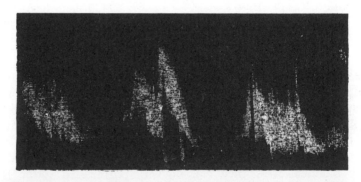

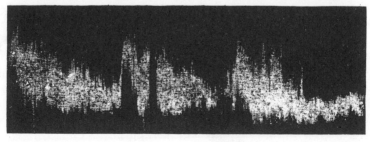

FIG. 10A

Curves made by ant activity in the apparatus described in the text. *Above*, a nycthermal rhythm may be noted, the lower part of the curve showing night activity. *Below*, no rhythm is recorded.

tive than was necessary: even though the ants only weigh two or three milligrams, more than one per second cross over and most of the time there are seven or eight insects on the platform at once; this is more than is needed to get an appreciable movement of the stylus. Straightforward observation suggests, and the recording on the drum confirms, that the workers pass in "gusts", as one might say, of 7 to 10 insects, separated by dead periods: this is why the balance swings without cease and the graph is not continuous. Enormous differences between tracks can be noticed, not only as regards general activity but also in respect of the daily rhythm. Paths not much used during the day are invariably empty at night; whilst the tracks active in daylight are also so at night, except during bad weather. . . . It is, on the other hand, impossible to establish any relationship between activity and temperature, *except at night*; then the correlation is perfect. The most noticeable thing in daylight is an important peak of activity about 10 a.m., which had already been noted by Wellenstein, by counting the prey carried back per hour.

Nest activity

One can go further than this by measuring, not the movement

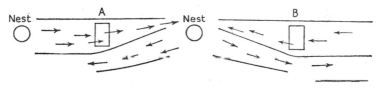

FIG. II

How ants reaching the recording apparatus are separated into those leaving the nest and those returning to it. The black lines indicate repellent barriers. Only the former follow the right-hand path and only the latter take the left.

on one particular track, but that of the whole nest. All that needs to be done is to surround it with a repellent barrier, as is done when using the automatic trap. But, instead of placing the perforated box on the route, I put two of my registering balances, one on the entrance path and the other on the out-going one. The apparatus has to be somewhat modified, because the number of ants passing at the same moment is

enormous and quite impossible to count with the naked eye:
also the balance arm should have a bit of lead wire (the weight
it had to be surprised me) attached to it to damp the over-
big oscillations of the pointer.

A most interesting phenomenon is then seen which is difficult
to prove by just watching the tracks: it is obvious that the ants
coming back from the forest should weigh more than those
leaving, since they will be swollen with honeydew or carrying
prey. And that is what is seen. The exit graph is notably lighter
than the entrance trace. Now if you collect the prey at the
same time, you can establish a ratio between honeydew and
"game". For instance, it has been found that much less
prey enters the nest at the end of August than at the end of
July. The graphs, however, still show an appreciable difference
between those leaving and those entering: this seems to be due
to honeydew. There is thus the possibility of assessing the parts
played respectively by the two kinds of food in making up the
total load.

Coefficient of food use

In any work on animal feeding you get nowhere if you only
consider what is eaten and *not what is rejected*. Ants are no
exception: they collect by hundreds miscellaneous pieces of
debris, composed of completely dried wings and legs of insects
and, moreover, it is impossible to believe they can assimilate
all of it.

I tried, then, to gather not only what they used but also
what they rejected. The solution seemed to be simple; it would
be enough to put the automatic trap in reverse, that is beneath
the exit passageway. In my cultures the ants very quickly
set up what we call "the cemetery"; that is to say a spot as far
outside the nest as possible to which they carry all the corpses,
cast skins of the larvae and food debris. I was thus led to believe
that they would do the same in the natural state and that all
debris would be left in my trap.

Not at all: I found nothing in the trap placed in this position,
or so little that it is not worth talking about; but on looking
closer I realized what had happened. I mentioned that the
nest was surrounded with a repellent wall, at a metre or there-
abouts away. Now against this wall, or not very far from it,

around nearly its whole circumference I saw a mass of whitened larval skins and other debris. The cleaners, it seems, *leave in all directions* and carry the debris away as far as possible. When I forced the outgoing ants to take a single road no doubt only the foragers consented to use it and not the cleaners. In order to collect what the cleaners took out, quite another sort of apparatus would be needed.

At the end of a few days I had done it. I just put round the heap a circle of continuous guttering, opening towards the top. Soon the cleaners had thrown a hundred kinds of debris, easily sorted and identified, into it. The results are quite surprising: over a month and a half I collected in this way: 6 kg. of fine sand, 273 g. animal debris, 268 g. vegetable debris and 23 g. larval skins; a few bodies of ants too, but in very small numbers. What is yet more surprising is that it was not by any means a big colony; on the contrary, it was small and had scarcely grown over two or three years.

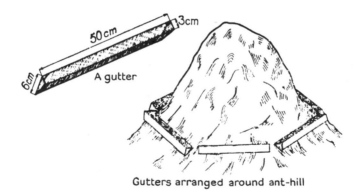

Gutters arranged around ant-hill

FIG. 12
Apparatus for collecting all detritus from nest. See text (after Chauvin).

Now, in the Parisian climate daily activity is only marked between 15 April and 15 September, say 150 days: this gives us for this period: 20 kg. of sand, more than 800 g. of animal debris, about as much vegetable remains and 78 g. of cast skins (say 500,000 hatchings if we accept 0·15 mg. per skin). Consequently this little nest was tremendously fecund and sets one thinking about the figures a big colony might produce; it is all the more remarkable, as Lange has pointed out, in view of the fact that the workers of *polyctena* can live for three years, which is somewhat exceptional for an insect.

But other facts also emerge from examining the animal debris: it includes woodlice and weevils which seem to be quite intact: the ants have scarcely done more than nibble their legs; enormous quantities of carapace remains are found. In the vegetable debris are a number of buds and seeds which the workers have collected in large numbers but with which they seem to do nothing. A Rumanian writer assures us that they sometimes cut off a considerable number of buds in order to damage the trees; and he adds that the ants feed on them. But neither one nor the other of these statements seems at all likely to me. Moreover, I do not see what food value the ants could get out of those buds.

All the vegetable debris has something in common: it is all made of short or round bits and excludes the long twigs used by the ants to build up the mound. From this I drew a conclusion: the foragers are driven by one simple compulsion, to bring back to the nest anything movable, alive or dead, with no other specification. *It is only later that the teams of sorters keep what is useful and reject the rest.*

Finally, by weighing the animal debris one could, with a simple calculation, establish the food use coefficient or the difference between what is brought in and what is taken out. I have not yet done it; at first sight this coefficient seems rather variable. In addition there is a relationship between excavating activities, shown by the sand rejection, and animal debris and cast skins put out; but there is little relationship between the whole activity and weather conditions.

The value of ant foraging from the point of view of the protection of forests from pests

I do not want to inflict too arid a lesson in entomology on the reader who may have followed me as far as this, but all the same, after this intense theoretical reasoning, it is time we turned to the views of specialists in agricultural entomology, who, as far as red ants and the forest go, are nearly all Germans. A short while ago the Germans passed on their myrmecological

fever to the Italians, who then put out some startling statistics. . .
From the estimated million nests in the Italian Alps, the weight
of prey brought in during the 200 days of annual activity
amounts to about *14,500 tons*, which provides food for an ant
population calculated at *300,000 million individuals*!

The importance of hunting by ants can no longer be doubted.
There is so much of it that one might even wonder if birds may
not suffer from such activities, for they too are our indispensable allies. Careful observations have led to one of the innumerable surprises of ecology: *bird population rises* in the woods
where ant colonies have been augmented, and we are not
speaking of ant-eating birds that increase. . . . On the other
hand, the number of colonies established naturally, with no
human intervention, does not seem to be correlated with the
number of birds. In spite of appearances their hunting methods,
or the prey they seek, must be different from those of the ants.

But, *do the ants destroy the parasites and predators of forest pests*,
that help us so much? There is no doubt about it; these useful
insects are found in the traps: according to Wellenstein again,
it is, above all, the insects which attack aphids that are up
against the ants, and among these the Ichneumonidae are most
numerous. We must add that the aphids attacking forest trees
are not of the kinds that kill the trees by injecting poison or
infect them with a disease; their effect on the health of the
tree is almost nil.

According to Otto, on the other hand, ants put into oak
forests at the rate of six nests per hectare have protected them
from the attacks of the leaf roller *Tortrix viridana;* they also
fight as efficiently against *Lymantria* and a number of forest-
damaging *Tenthredinidae*. Sometimes, even if they do not directly
attack the pests, their comings and goings disturb the development of the pest species. And the occasional jet of formic acid
shot at pests kills them stone dead. The problem is simply the
number of nests one should put in and it varies, according to
different authors, between four and ten per hectare.

CHAPTER II

OTHER, MORE OR LESS LIKEABLE ANTS

Migrating warrior ants — Agricultural ants — Harvesting ants — The fire ant, an enemy of man

A FEW YEARS ago cinemas were showing a horror film of red regiments of ants destroying a farm and devouring its inhabitants. The film had been a little "faked": it gilded, if I may say so, the lily of reality. A man in good health has nothing to fear from warrior ants: all he needs to do is to get out of their way. A sick or wounded man might not be so fortunate and could well be eaten. And I have been told that it was a favourite diversion of the Baoul kings: they tied down on the ants' track people they did not like and a week later only a nicely cleaned skeleton remained; the first quarter of an hour of the attack must have been horrible. We must point out that, according to Raignier, a single colony of *Anomma wilwerthi* can contain more than 20 million ants!

The point here is that we are dealing with a particular species of ant having no fixed nest, but only temporary "camps". When they stop their marching the ants form a large ball containing the queen and the brood (70 cm. high and 60 cm. in diameter!). They may stay like this for a few hours or several days; then they start off again, forming a column, and enter the forest in search of prey. I have several times found myself face to face with these hunters in Africa: the big soldiers, with enormous jaws, stand immobile at various points of the black column of workers (the African driver ants are black), about the thickness of a finger. By touching each other the soldiers' pincers form a kind of dome under which the workers pass at a speed of about a hundred to two hundred metres an hour. They can pass by like this for hours and hours on end, so that if the soil is friable they may hollow out a trench five to six centimetres deep. The ants can be watched without too much risk provided you do not get too near. If you approach and cross some invisible line, which must be about 25 cm. from the column, the soldiers become active and come towards you; or you may notice nothing until a few stinging bites suddenly let you know that a squad of workers has turned your flank and attacked from the rear.

An American, Schneirla, has made himself the expert on the army ants and has solved the mystery of their migrations. I have taken from him a large part of what I am now going to tell you. These warriors belong to the *Dorylinae* and there is some palaeontological evidence leading us to believe that they come down to us from the Tertiary era. . . . Thus, long before man, the late arrival, these fierce columns, red and black, tracked through the forests; and they have scarcely changed since then, any more than other ants, as we saw at the start of this book. The African species belong to *Anomma* and the American to *Eciton* and they both inhabit tropical forests. I admire the stubbornness of Schneirla who, somewhere or other, describes the frantic gymnastics he had to perform time after time in order to follow the column through the twisting creepers of the underwood; and how many times a false step precipitated a shower of furious *Eciton* down his neck or into his boots. I much sympathize with him over these troubles, because I tried to do the same thing in the Ivory Coast and they soon cooked my goose. . . . The most impressive forays are by a big species, *Eciton burchelli*, which starts off by sending out a close-packed column some 15 metres wide and two or three metres deep. Then a preferred direction becomes established little by little so that a kind of tree with many branches is formed.

Generally one can detect the presence of the ants, not by seeing, but by hearing them, the more easily in that the virgin forest is not noisy in full daylight. It is only a little after sunset that the pandemonium of cries and whistling arises, which swamps the noise of the warriors passing. But in daylight, or rather during the half-light always found even at midday in the primary forest, one may hear a series of patterings and crackling not unlike the sound of a shower of rain: it is the sound of the innumerable victims falling from the trees, chased off by the

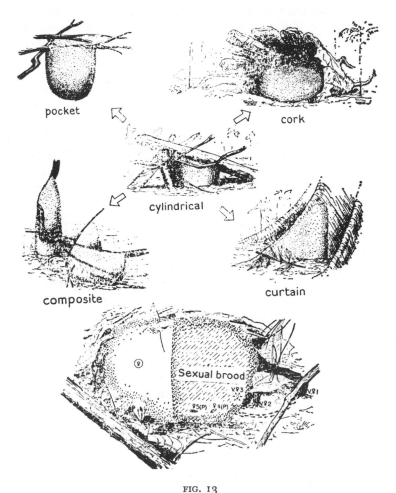

FIG. 13

Five upper figures: Different forms taken by the temporary nests of *Eciton hamatum. Bottom:* Section of nests at the time of hatching of the queens. V♀1, the first to hatch, is 50 cm. to the right surrounded by a handful of workers. V♀2, the second to hatch, again surrounded by a few workers. These two queens continue to be attached to the nest by a column of workers carrying food. V♀3, young queen still in nest. ♀4 and ♀5, royal pupae. On the left is the still active actual queen (after Schneirla, 1966).

ants, whilst another body of workers waits to dissect them in the green depths of the forest. If now you approach them, with suitable precautions, from a little distance away you will notice a pronounced buzzing: it is caused by a swarm of flies hovering at a fixed height above them; the flies are trying to lay eggs on the ants' bodies. The tables are then turned, for the fierce warriors will be eaten alive by the maggots that emerge from these eggs. For these migrant ants have not developed an instinctive knowledge of how to rid themselves of the parasite's eggs, although it has recently been discovered, as we shall see, that the fungus-growing ants manage to do so.

Sometimes the buzzing of the flies is drowned by the shrill cries of countless birds. The old authors thought the birds were feeding on the ants, but this is not the case: they are happy to gobble up some of the victims the ants have chased off branches, catching them in falling, one might say under the very noses of the ants if in fact they had noses! Indeed, very few vertebrates attack army ants. It is not because there are no ant-eaters in the tropical forest; Schneirla has often seen ant-eaters and coatis there, whose favourite dish is ants. But not these. The large animals seem definitely to be afraid of the soldier ants' terrible jaws; so much so that, no doubt, it is left to the flies to prevent an impossible increase of *Eciton* and *Anomma*.

What do the warriors catch? The big species, such as *Eciton burchelli*, which even attack small mammals and reptiles, take almost anything found in the forest, whilst the smaller kinds restrict themselves to invertebrates. According to Raignier, in one night a single *Anomma wilwerthi* colony can eat some ten chickens, five or six rabbits and a sheep, and even that is not an exceptional bag for such a redoubtable species. As well as these anything they can find will do: tarantulas, scorpions, spiders, cockroaches, as well as the brood and adults of many other species of ants. Not all, however, for there are some kinds who can stand up to the army ants, the fungus-eaters for instance. Also, in Africa, *Œcophylla*, or weaver ants, successfully face up to *Anomma*. In any case the wild is empty after they have passed by; one expedition can accumulate *more than four litres* of insects.

Their voracious appetite can even be used by man. In tropical Africa, when I used to sleep practically in the open air, sheltered only by a thatched roof, a mass of somewhat suspect life woke in the thatch every evening to such an extent that I found it hard to go to sleep even after a day's tiring work. . . . It was ceaselessly humming, stridulating, creeping and scratching. I know the mosquito net shelters one to some extent from the swarming life of the African hothouse; but when *Anomma* is on the warpath, well. . . . For a few hours it is wise to make way for them: they will leave nothing alive and nothing undesirable in the place for, as cleaners, they are irreplaceable.

If you watch the migrants on the move, several streams going in opposite directions may be seen: some carry nothing, the others, moving the opposite way, are heavily burdened: they are going back to their camp. Later we will see how this is organized. What first strikes one is their exaggerated difference in size: with *Anomma* the smallest workers only measure 0·5 mm. whilst the largest reach 14 mm.; as to the males, they exceed 30 mm. Finally the enormous queen reaches 60 mm. and her abdomen is 16,500 times bigger than that of her smallest worker! But if you look closer, under the flies trying to lay eggs on the ants, you can also see a strange host of small or minute insects, humped, hairy, some resembling ants, others clinging to the workers themselves, which seem to tolerate them: these are what Wasmann called "symphilids", in other words their commensals, some of whom do not entirely benefit the ants. Most of the time these uninvited guests actually exploit their warrior hosts quite shamelessly without suffering for it. But nothing is known of their habits, at least with *Anomma*. To find out you would have to get near to their terrible bivouac, and that is not a risk I would care to take. I once saw two of our African workers by mistake plunge a mattock into a colony: a crawling mass of ants instantly covered them up to the knees and they executed some very lively dance steps before taking to their heels. . . .

The "bivouac" is a ball of ants, often too big to get into a bucket; the ball is usually situated a little above ground, though some species of warrior ants have a nest below the surface. The workers cling to each other and are almost immobile, but they leave a few passages and galleries open in the ball. The sole queen, huge and almost blind, is found towards the centre, with the impressive brood—60,000 larvae and pupae at times in the case of *Eciton burchelli*; with *Anomma wilwerthi*, according to Raignier, the brood can fill three buckets or more.

As in the case of bees, the assembly of such a large number of insects gives rise to a "microclimate". Inside the ball of ants the movements of the external temperature are much damped and the inside can often be 12°C higher than the outside temperature, above all in the brood region, especially during its development stage, until the larvae spin their cocoons. At times the ball is nearly stable and the daily raids are not very frequent and do not go far; on the other hand, at other times the ball is

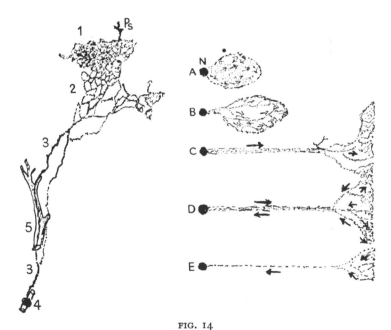

FIG. 14

Left: Plan of an *Eciton burchelli* sortie, after five hours' travelling. 1. The major part of the swarm has gone out 70 metres on a 15-metre front. At P_S, a branch movement leading to a new advance; 2. Fan-shaped advance; 3. Main column; 4. Bivouac nest, inside a rotten stump; 5. Remains of a tree trunk on the ground, to the right of this a secondary column has been formed by the division of the main column (after Schneirla, 1966).

Right: Development of an *Anomma wilwerthi* sortie. A. The swarm leaves the nest; B. The swarm breaks up into branch columns; C. The sortie at its height; D. The return to the nest starts; E. The return continues (after Raignier and von Boven, 1954).

remade almost completely every evening, even being reformed elsewhere.

This is the problem that interested Schneirla. . . . What was the cause of this marked variation in these ants' aggressiveness? He admitted that at first he was inclined to the view that the weather controlled them, more especially as the sorties were very regular: sixteen days for the normal (hunting) phase and twenty days for the quiescent (stable) phase. That is easily related to the moon's phases. At the same time it was easily seen that several other colonies in the same area were at different stages of the migration cycle, which ruled out any external factor working uniformly.

Various details then put the American worker on the right track, quite a different explanation: for instance, the fact that at the end of the quiescent or static phase, a cloud of open cocoons fell to the ground near the bivouac, and at the same time many young workers could be seen, recognized by their lighter colour. On the other hand, one can easily see that in the nomadic phase workers carry numerous larvae in all stages of development, but no pupae. One thus concludes that in some way or other the ants' activity is related to the age of the brood. And the state of the brood obviously depends on the queen. On the island of Barro Colorado, where he made his observations, Schneirla easily managed to find the queen (one wonders how he could do it without being eaten alive!) Now, during the migratory phase the queen manages to move, at least at night, in spite of her enormous size. Schneirla says that her approach is heralded by a sudden increase in the excitement of the workers in the column. During the static phase the queen's abdomen is tremendously swollen, to twice its normal size, by the eggs: it is at this time that she lays 60,000 or more eggs, at times in less than a week. A little later, when the young larvae have hatched, activity starts again; the workers start the trekking phase again, and the queen can follow, her abdomen, now free of eggs, being of normal size.

What happens is as follows: the presence of active larvae wanting food greatly stimulates the workers (which can be proved in the laboratory); they are induced to hunt continuously in order to satisfy several tens of thousands of ravenous appetites: this is the nomadic phase. But as the larvae approach pupal stage their voracity falls off until they stop eating. Hunting activity does not stop so suddenly. No doubt this is the nub of the matter: as the hunt goes on for a while and the victims have no other consumer but the queen (that is, there is no other consumer save the queen for the enormous surplus remaining after the workers have satisfied themselves) she stuffs herself with food. Her abdomen starts to swell enormously, her food reserves build up and her ovaries start to function. At this time she is incapable of movement and also at this same time the larvae have spun or are spinning their cocoons; it is the start of the quiescent phase, which ends with the hatching of the pupae and the laying of the eggs.

FIG. 15
A soldier of the driver ant (*Anomma*) on guard on the outskirts of a colony (from a photograph by Raignier).

The nomadic ants make new queens at irregular intervals, often separated by long periods of time. This usually happens after a dry spell. There is not much brood fed to produce the sexuals; no doubt this is explained by the mechanical difficulties of sustaining the enormous larvae and pupae. Whilst these are developing the colony can scarcely move and the camp is more or less fixed. Of the five or six queens being raised, only one will be kept: she will be without wings whilst the equally big males will have them. They are thus easily seen, especially at night

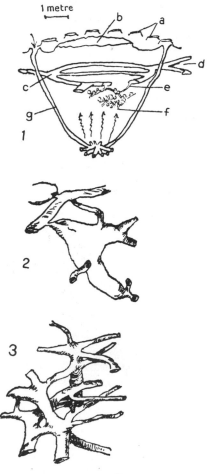

FIG. 16

1. Nest of *Atta sexdens* (drawing by Sudd, from Jacoby's observations, 1955). (a) crater entrances; (b) ring of excavated soil; (c) circular tunnel; (d) entrance tunnel; (e) sloping tunnels with fungus gardens f; (g) outer tunnels on nest margins. The wavy arrows show the supposed air circulation.

2. Enlargements in the tunnels, called barracks by Jacoby.

3. Big "road flyover".

when using a lamp. On the Ivory Coast they have made many nights miserable for me although they are inoffensive enough. These winged ants, falling by dozens into your hair or plate, are not very tempting. Their considerable number is probably due to the fact that, as with bees, and *Atta* ants, the queen needs plenty of sperm to fertilize the eggs throughout her long life. Consequently she has to mate a considerable number of times.

As soon as the young queen has hatched the colony separates into two portions. In some cases (as with *Neivamyrmex*) all the males leave with the party of the still unmated young queen. In other cases, as with *Eciton*, the males are divided almost equally between the two halves of the old colony.

The fungus-eaters

There are some ants, species of *Atta*, who are agriculturalists, or, more precisely, fungus-cultivators; and I am not speaking metaphorically: it is just that, as we shall see. These ants are remarkable not only for this behaviour, but for other habits as well. For instance, their underground nests are without doubt the largest there are among ants; Autuori estimated that *22 cubic metres* of earth had been moved by workers in a single nest. The ants' constructions are only excelled by those of the termites, whose nests, such as those of *Bellicositermes*, enormous and a hundred years old, are more than twenty metres in diameter; at times African villages are built on them.

Fantastic differences in size and shape (polymorphism) are found among these fungus-eating ants, some idea of which is given by Fig. 17.

It should also be noted that *Atta* are not the only leaf cutters: and there are other, much less well-known species who grow fungi, but on insect excrement or on fine vegetable debris. The *Atta* prefer leaves, which they cut to such an extent and so often that they can defoliate an orchard in a short time. Thus they are classed as pests and ones, moreover, that are very difficult to control, in view of the size, population and dimensions of the nests.

Nevertheless, *Atta* may play a useful, though unexpected, part in agriculture. Their huge nests in fact mix in a considerable quantity of organic matter to a great depth in the soil.

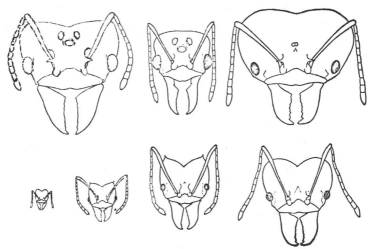

FIG. 17

Heads of the different castes of *Atta cephalotes*, drawn to the same scale.
Length with mandibles given first, then distance between eyes in
brackets mm.
Top Row: Queen 5·8 (3·92); Male 4·1 (2·98); Soldier 5·8 (3·45).
Bottom Row: Minima worker 0·9 (0·58); *Media* worker 1·8 (1·16);
Another *Media* worker 3·5 (2·03); *Maxima* worker 4·3 (2·53). Body
lengths are: *Minima* 1·5–3·5; *Minores* 3·5–6; *Media* 6–9; *Majores*
9–13; Soldiers 13–16 (after Weber, 1966).

This is important in tropical soils, which are low in organic
material, contrary to what is popularly thought to be the case:
the hothouse temperature of those climates causes the gentle
bacterial breakdown, giving humus, to develop too quickly and
to go too far. Anything which might usefully feed other plants
is completely degraded and in addition torrential rains often
carry away decomposing matter. *Atta*, and all the fungus-ants,
help to set this right. Remember, moreover, that there are
areas where a nest may be found every two square metres.
It is a matter of common observation that trees and plants are
greener and more robust near the *Atta* nests.

These nests, and some have had *forty tons* of earth excavated
from them, contain thousands of chambers where finely minced
leaves are sown with the fungus. Autuori studied one of these
nests and counted 1,920 chambers of which 248 contained an
active fungus garden (vegetation and brood) weighing 300

gms.: say 100 kg. for the lot: according to Autuori, the in-
defatigable workers, during the active period of this nest, must
have carried down more than *five tons of leaves*. One easily sees
why they are considered a pest by farmers. The nest is not made
up of irregular cavities, as is that of the red ant, but is built to a
strange plan (see Fig. 16). One of its strangest characteristics
is the presence of *immense chambers a metre wide and more than a
metre high* which the ants use as a cemetery and refuse pit:
these are by far the biggest cavities ever made by an insect.

The giant *Atta* nests (see Fig. 16) have been studied by
Jacoby, who revealed the plan of construction by pouring
liquid cement into the openings. He was able to calculate the
interior volume of the nest by the amount of cement necessary,
getting the almost unbelievable figure of one-and-a-half cubic
metres. He found 120 metres of tunnels of an average width of
three cms. (900 litres), and 100 fungus gardens each with a
capacity of six-and-a-half litres. Outside the nest showed as a
dome of loose earth of from five to forty-five cubic metres,
surrounded by a series of craters. Underneath there was a
veritable network of more or less horizontal tunnels; below that
again were the fungus gardens. The craters are not used as
entrances. The supplies of leaves are brought in by horizontal
tunnels, the openings to which often lie far beyond the craters
and central disc, leading into a large central gallery where the
leaves are stored for some time before being transferred to the
fungus gardens via a slanting tunnel. As to the craters, they
open into roughly vertical galleries that surround the nest,
and converge towards its base, but have no direct communication with the fungus gardens. The temperature inside these
nests is from about 25° to 28°C and may rise to 29°C in the
gardens themselves, exceeding the external temperature of the
surrounding earth by 15° or more. The concentration of carbon
dioxide gas in the galleries' atmosphere must reach four times
that of the outside air, with a humidity approaching saturation
point.

Obviously this raises the problem of *nest ventilation*, as in the
termitaries, where conditions are very much the same: enormous volume, huge population and a great quantity of organic
matter in full fermentation. Lüscher maintains that with the
termites a complicated system of chimneys produces a reasonable

amount of draught; but not everyone accepts his theories. Be that as it may, the air in an *Atta* nest must renew itself to some extent or the heat produced by fermentation would hardly be dispersed at all, and the temperature inside the nest would rise unchecked as in those straw stacks that have been known to burst spontaneously into flames. As yet an accurate study of the ventilation and heat control of an *Atta* nest remains to be done.

Obviously the centre of interest is the fungus garden itself. Some species, such as *Acromyrmex*, make only one, measuring some forty to fifty centimetres in diameter. Those of *Atta* are very numerous, with a diameter of twenty to thirty centimetres. The ants bring in from outside fragments of leaves, mince them up into ever finer shreds and water them with a drop of anal liquid: only then are the leaves incorporated into the hotbed. After that the ant detaches a fragment of fungus mycelium and sticks it on to the new layers.

The ants never subsequently abandon the fungus garden. They traverse it in all directions, brushing it with their antennae, licking—and sometimes eating—the hyphae. But above all, and this is the most curious biological problem set by *Atta*— *they keep the culture pure.* In spite of the presence of countless spores of bacteria, of yeast and other fungi, only one species of fungus grows on the beds, and that one is characteristic of each kind of ant. As Weber observes, this implies that the ants produce several factors: a fungistatic substance that inhibits the growth of any fungus except the one cultivated by the ants; a *bacteriostatic* material, and probably also a growth regulatory substance.

These substances have never been properly studied, however. This is due to chemists and pharmacologists being interested almost exclusively in plant products and having nothing but contempt for the old pharmacopeia, which made great use of drugs of animal origin. Nevertheless, animals, and especially insects, secrete all sorts of substances, some of them very extraordinary. We will see various examples below, such as poisons secreted by ants and recognition and marking odours. But the importance of bacteriostatic material, and above all of a fungistatic substance, is quite otherwise: who knows if one day some important medical use might not be found for them?

The fungus cultivated by *Atta* has been the subject of much dispute. As we have seen each species of ant has its own particular fungus and cannot change it. The fungus, on the other hand, seems equally tied to the ant, because it is never found anywhere else. Apart from that, our knowledge of how the fungus is grown is still very vague; for instance, we know nothing of how the fungistatic substances are spread, nor their source in the ant's body: are they mixed directly into the leaf bed? If yes, then all the time or at intervals? Do all the ants secrete these substances? This last is not likely, because one kind of ant, the very small workers, the *minimae*, are especially tied to the fungus gardens, to such an extent that you cannot break off a piece without at the same time taking some *minimae* tightly clinging to it: so that perhaps they alone are the excreters of the fungistatic substance.

As I said at the beginning, an important point about the large leaf-cutting species (*Atta* and *Acromyrmex*) is their fantastic

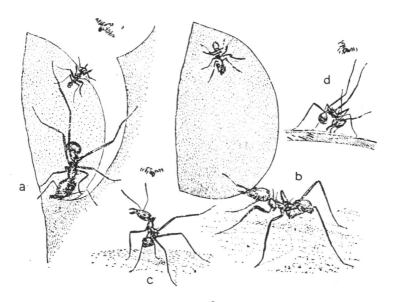

FIG. 18
Atta workers with a parasitic fly: a. A *media* worker cutting a leaf whilst a *minima* watches the fly; b. The *minima* remains on the leaf as it is carried away; c. Alarm position of *minima* during the attack (after Kacher, in Eibl-Eibesfeld, 1967).

polymorphism of which Fig. 17 gives an idea. The largest individuals are the soldiers, having enormous jaws which can easily penetrate not only human skin but shoe leather as well. They stay inside the nest, near the queen, and defend the colony. The middle-sized workers cut the leaves from the plants and the *minores* and *minimae* chew these up to make the fungus bed. The middle-sized workers also look after the brood. In 1967 Eibl-Eiberfeld found that the *minimae* took on another important special duty. Many authors have noted that a piece of leaf carried along by the big workers often had on it a very excited, tiny worker, its mandibles open "as if threatening the heavens". And, in fact, the danger could come from the skies, in the shape of a small fly, a Phorid (*Apocephalus*), trying to settle on the back of a large worker's head in order to lay an egg: the larva coming out of this egg would eat out the contents of the ant's cranium. The tiny ants are a defence against these flies on the return journey, when the large workers are too intent on their burden to have time to defend themselves. If, in spite of all, the fly succeeds in depositing its egg the *minima* hurries to lick the neck and head of its companion, thereby no doubt removing the parasite's egg. . . . The queen is enormous, and is perhaps, with termite queens, the insect with the longest life: twenty years is spoken of, a period during which the sperm, stored once and for all at mating time, has sufficed to provide for the renewal of all the workers in the colony. A count of the spermatozoa present in the seminal vesicle gives most surprising figures, of the order of several millions, which shows that *Atta* queens, like queen bees, mate several times during the nuptial flight. The queen has a particular structure peculiar to ants, but in this case especially adapted to fungus culture—the *infrabuccal sac*, a hollow situated at the back of the mouth cavity. A mass of debris cleaned off the skin, by means of a special comb on the ends of the tarsi, collects there. Ants are somewhat hairy and their skin retains a whole series of germs and dust: the cleaning process, carried out so meticulously by insects, removes all that. But with *Atta* the numerous spores of the fungus necessarily accumulate in the infrabuccal sac, as well, of course, as many other kinds of spores; it does not matter as the bactericides and fungistatic substances will soon inhibit the latter.

The queen makes her nuptial flight, then sets out to found a new colony. She digs out a hollow in the ground and does not move for several days, She starts by laying a special kind of egg, called "food eggs" by Autuori, because they are not intended to reproduce the species: they are rounder and bigger than the normal eggs. The queen crushes them and spills out the contents of her infrabuccal sac on them. The mycelium starts to grow and the queen will allow no descendants to be born until the fungus is growing well. Then the first workers will in due course go out looking for leaves with which to make the fungus gardens. . . .

Is such a strange way of life an isolated case in the seething world of insects? Not at all. Though *Atta* possibly has best exploited fungus cultivation *it is not the only species to do it*; for instance, termites are well aware of it. But their fungus is not the same and it grows on pulped wood reduced to a spongy mass. Even so the termites do not eat the fungus itself but the lower part of the woody mass which has been made assimilable by the fermenting action of the fungus; even lignin, a substance almost as stable as paraffin wax, is broken down to simpler substances on which the termites can feed. They too have a pure culture, consequently they also must excrete bactericides and fungistatic substances; and here too we know little of how they cultivate their fungus. Be that as it may, I shall not forget my encounter with fungus-growing termites in the forests of Gabon. In the heart of that humid, green twilight stood the yellowish termites' nest, waist-high; it was of a consistency soft enough to slice with our machetes; disclosed lay rounded chambers each holding a sort of brown sponge speckled with white by the fungus. It was the strangest of spectacles. But the termites had no means of defence and let us do our worst. In contrast the first time I came up against *Atta* the soldiers made me regret it.

How did the ants become fungus growers?

In order to understand how such a strange way of life could arise in the course of evolution we must, as always in biology, look round the whole world of ants and enquire into the extent of fungus cultivation. It is not at all common. We know, for instance, that in the tropics moulds easily grow on the remains of the tree nests of *Crematogaster*; but *Crematogaster* has never been seen to take the slightest interest in them. Many plants are known as "myrmecophytes" because they have natural or excavated cavities sheltering certain kinds of ants, always the same species; and inside these cavities with a saturated humidity a number of fungi grow. Ants have even been seen to cut or pull them off, but proof that they are eaten has not been obtained; perhaps the ants were simply cleaning the nest. In addition, Weber has justly remarked that typical *fungus-growing ants are never found in the cavities of American myrmecophytes*, whose dimensions might suit them very well; they prefer to set themselves up alongside under very different conditions.

It has been known for a long time that the blackish paper-like substance provided by one of our temperate zone ants (*Dendrolasius*) shelters a special fungus (*Cladosporium myrmecophilum*). But we know nothing of what the ant does with this, nor if it is eaten.

With true fungus-growers a special habit may be the origin of cultivation: that of defecating always in the same place. On the mass of the colony's semi-liquid faeces it is very likely that some fungus or other could develop, and not just any old one, for the kind of growth possible is dependent on two factors: one, the chemical nature of the excrement, and two, the habit of cropping the fungus closely and constantly. Only a species with a luxuriant growth could survive under such conditions.

A second stage would be to enrich the site with a variety of substances of which the first might well have been the excrement of other insects, for instance of wood-boring beetles or of caterpillars; in this last case vegetable particles such as undigested bits of leaves would be found in the material. From that the ants would go on to fallen, dead leaves; and then to the final stage of cutting off and mincing up green leaves.

Now the above is not just the result of a lively imagination, because these different stages can actually be found in other fungus-growing species. In the most primitive species, *Cyphomyrmex*, the workers prepare the fungus bed, to wit the excrement of other insects, by licking it and by adding their own dejections. With higher species, such as *Acromyrmex* and *Atta*, the fragments are licked, nibbled at the edges, and very often the ant drops excrement on them; only then is the fragment put on the bed, the ant using its mandibles and—which is rare among insects—its front legs.

Other agricultural ants: harvesters

Messor of North Africa dig enormous nests, seven to ten metres in diameter, and as deep as two metres in the soil. They can be found thanks to the presence on the surface of several craters about 50 cms. in diameter. The storage chambers are flat cavities of about 15 cm. broad by 1·5 cm. high, joined to each other and the surface by a number of galleries. *Messor* is not the only ant to harvest grain; quite the reverse, the habit is often found among ants, for example *Pheidole* and *Pogonomyrmex* (from the Greek "bearded ant") of America, whose crater is made of grit and is at times 50 cm. high and whose enormous nest can go down as far as five metres.

A particular habit of these harvester ants is to make long paths, sometimes of 100 metres, very wide and cleared of all vegetation. The fungus-growing *Atta* do the same thing; one would swear the paths had been made by man, and somewhere Gœtsch describes the astonishment of lost travellers who have followed one such path, hoping to come to a human dwelling; and all they found at the end of the road was a huge ants' nest, active, hostile and inhuman.

Here is a good place to mention a theory put out by Lincecum from 1862 onwards, which caused a considerable sensation: this biologist maintained that the harvester ants cultivated cereals around the nest. Now, it is true that often one finds a crown of cereals, at times of a single species, around a nest, which suggests the idea of deliberate cultivation. Remember that this would not be particularly surprising; *Atta* cultivates fungi and you will see, or you have already noted from this book, that ants are capable of anything. However, in the case of these harvester ants Wheeler showed how the mistake arose: from time to time the workers bring out part of their store and put it in the sun to dry, particularly grains that have started to germinate. If rain falls the ants abandon them and the grains are then free to grow. Moreover, as Wheeler correctly maintained, the enormous roads that leave the nest show that a cereal garden of a few thousand plants would not suffice to supply the needs of the harvester colonies.

Another singularity is the presence of the somewhat badly named *soldiers*, because they are not in the least belligerent: they are less aggressive than the workers and their enormous mandibles know not war: in place of that they serve as the colony's "nut crackers", as Wheeler puts it: it is they who break the grains and cut up the largest and hardest of the animal victims brought in. Thus the "soldiers" are very useful, but not in the way one might suppose by their having been dressed up with such a name.

I mentioned that many species of ants collect grains as well as *Messor*. A still greater number collect them occasionally: *Tetramorium caespitum, Myrmica rubra, M. brevicornis, Lasius fuliginosus, L. niger, F. exsecta* and *F. polyctena*. But what exactly do they do with them? Do not let us forget that we are dealing with an insect where the strangest collective results accompany the most ridiculous individual "mistakes". With *polyctena* I am pretty sure that the collecting of grains is no more than a part of the collecting frenzy of the workers: for instance, I have been able to observe (see page 66) that a great number, if not all of them, are rejected without having otherwise been touched. It seems that what interests the ants is not the grain itself but its oily coating, at times distinguished by a special appendage. The proof is that if you coat glass beads with a vegetable oil the ants readily collect them (at least according to Steiger; but I myself have seen *polyctena* collect and carry back to the nest numerous little pebbles without their having been coated with anything at all). However, *Messor* and *Pheidole* do really eat the grains they collect.

Gold-seeking ants

You may rest assured that ants give no special importance to gold or any other metal: however, they do help prospectors of gold and rare metals. Gœtsch found this out in the case of a species of *Dorymyrmex*, adapted to the desert, and living in the desolate steppe of Atacama, Chile, an area rich in many minerals. Like all desert ants *Dorymyrmex* is obliged by the drought to push its galleries to several metres' depth in the soil; the spoil forms an easily visible hump in which one can often find grains of gold or granules of copper ore; manganese ore in the New Mexico desert. Prospectors in search of the least sign have known the value of these ants for a long time. Moreover, Herodotus had heard it said that ants would look for gold: obviously they were not looking for anything for, to them, it was just useless material to be pushed out of the nest.

The fire ant (Solenopsis saevissima)

Not many years ago the farmers of Texas saw a large part of their fields erupt, almost overnight, with rounded anthills from which, at the slightest disturbance, poured light-coloured ants with a very painful sting, capable of causing serious damage. After getting over this unpleasant surprise the Texans philosophically said, "Well, another ant then; a drop of insecticide in the nest and they'll be gone. . . ." A few years later, in spite of liberal applications of various poisons, the ants were as bad as ever and the pest continued to extend its zone of action. Why? How could an ant defy the powerful means of control we have today? There is a complex biological answer to this question.

The fire-ant invasion

The commencement of the invasion was slow and unnoticed, all the more so because there are three indigenous species of Solenopsis in the area with nothing remarkable about them. A black form of S. saevissima (Latin: "the very cruel") was imported by mistake from South America in 1918 and remained in the neighbourhood of the village of Mobile for some ten years, where the slight damage it did remained almost unnoticed.

But all changed when a new Solenopsis appeared, a light, reddish brown colour, smaller than the black ant and with a smaller nest. This was in 1930 and the new kind must have come from the north of the Argentine or from Bolivia. It spread out from a focal point and drove out the black form, which is now rare and in danger of becoming extinct. Apart from its sting, one of the principal kinds of damage done by the ant is due to its voracious appetite; not only does it eat vast quantities of food and household products, but it also seems to have still further extended the range of its foods: not only insects of all kinds, cereals (above all maize) and groundnuts, but also chickens or even calves or piglets which, if they do not devour them, they at least wound seriously or kill. They have some characteristics which are quite rare in the ant world: they not only store grain (a habit also found among the different kinds of harvester ants) *but also remains of their animal victims,* such as dead insects. *This is unique* among ants, and perhaps among social insects as a whole. (It is claimed that the *Polybiinae,* social wasps of South America, store their dead or paralysed prey, but up to now this has not been confirmed, as far as I have heard). In any case, the Hays, while examining *Solenopsis* nests, have found hoards of dead and living termites, as well as locusts, in the special galleries.

Another drawback of the fire ant is the large number of its nests and heaps (up to fifty per acre) which render the ground so rugged and uneven that it becomes very difficult to work because of damage to farming machinery. Owing to that, and because of the worker ants' ferocity, it has sometimes happened that farm labourers have refused to work in certain fields, so that the owner has had to abandon them to the ants. And that is not something that occurred in the Middle Ages, but in our own time, 1965 to be exact.

Solenopsis baffles the poisoners

And so the above-mentioned Texan farmer's reaction, "Spray everything with insecticide and say no more about it", seems natural enough. But we are not dealing here with ordinary insects, though their powers of adaptation can be disastrous enough, but with a social animal on the top rung of the ladder of insect psychology. And that is where the plot thickens.

They began by spraying, under pressure and inside the ant-heap, a fairly strong dose of a powerful insecticide, dieldrin, which well and truly exterminates *Solenopsis in the laboratory.* But *under natural conditions* results are much less satisfactory. Wilson discovered the reason by excavating round the heaps. As soon as spraying started, the nurses carried the brood away through a maze of underground passages and placed it in safety at some distance; and it was not difficult to see that more often than not it had suffered no damage. The larvae can survive in the laboratory just as long as those taken from un-molested nests. And soon afterwards the heap will be back again in exactly the same place.

This treatment has, moreover, a grave disadvantage: dieldrin is a fiercely toxic substance and not only to insects, for a pound of it to an acre of marshland will destroy thirty sorts of fish and all crustaceans; only the molluscs survive. A similar treatment might be tried with fumigants, whose action is much quicker. But the results are no better, probably because the nest is too large and not permeable enough to the gas, and once again the greater part of the brood escapes. And—peak of perversity—the workers return to the nest afterwards and enlarge the openings as if they sought to air it. A few days later the whole tribe is triumphantly reinstalled.

But we must not give up so easily: I mentioned just now the voracious appetite of *Solenopsis*: how would it be if we offered a succulent but deadly dish? Indeed, poison baits are not entirely ineffective. But the young fertilized queens that burrow into the soil before, during or after one of the above treatments remain underground without eating, and touch absolutely nothing. By the time they have produced their first workers the poison has long lost its strength, or been carried away by rain. The next idea was to combine the scattering of poison bait with surface sprays of insecticide so that any queen that escaped one operation would fall victim to the other. However, the insecticides used, chlordane and heptachlor, although generally very efficacious against ants, *are repellent to Solenopsis*! So we fall between Charybdis and Scylla: after the spraying most of the huntresses stay in the nest, and no one touches the bait. Obviously we must find another insecticide, one that is not repulsive. Only we begin to wonder in what way those diabolical *Solenopsis* will parry our next thrust!

So what is left for us to do? To be precise, what we should have done at the beginning: make a thorough study of the ant's biology in order to discover a loophole through which we can attack it, some weak spot in that baffling armour of behaviour—very simple, no doubt, when each element is considered in isolation, but which has constituted up to now an unbreakable system of defence.

That is indeed what Wilson has been attempting in studying the "chemical language" of *Solenopsis*. But the path that will lead us towards a definitive solution of the problem still stretches a long way before us.

CHAPTER III

THE ANTS' HERDS

Many species of ant find aphids and scale insects attractive, because these insects exude through the anus the surplus of the digested sap they have absorbed. It is a liquid fairly rich in sugar, very poor in nitrogen, that the workers seem to crave quite desperately. This is not difficult to verify: in summer if you notice a colony of aphids anywhere, it is rare indeed not to find an ant "milking" them, that is to say patting the end of the abdomen with its antennae tips to make the aphid exude a drop of sugary excrement. But there is more to it than that, as we shall see, to the point where it is reasonable to describe the aphid-milking ants as stock raisers, just as one can speak of agriculture in connection with the fungus-growing ants.

There are indeed aphids, such as *Forda formicaria*, that are found only in ant nests. But the ants themselves are very eclectic in their tastes; the red ants, for instance, keep sixty-five different species of aphid.

The famous myrmecologist, Wheeler, considered that during the course of thousands of millennia ants and aphids have become biologically adapted to one another. In truth, the aphids make no attempt to evade the workers, although they do not lack weapons of defence, namely cornicles situated at the rear that, in case of attack, spurt out a sticky substance. It is pretty efficacious and, should ants attack aphids belonging to species other than the ones they herd, they are liable to get their heads and jaws glued up and retreat in disorder. Now, *with aphid species herded by ants the cornicles are often reduced in size or even entirely absent.* These aphids exude the honeydew gently when the ants ask for it, instead of ejecting it to some distance as they do when their milkmaids are absent. More obligingly still, some have developed near the anal orifice a circle of hairs that retain the honeydew till the ants come to fetch it.

These curious adaptations have not been observed in scale insects although ants frequent them just as assiduously. It

FIG. 19

The head of an ant (*right*) compared with the hindquarters of an aphid, whose hindlegs resemble antennae (after Klort, 1959).

may be, however, that the relationship between ants and these insects has not been studied quite so closely.

On their side, Wheeler reports, the ants return the aphids' courtesy first of all by not killing or eating them(which is only true in a general way). Further, they do not approach them aggressively, but with a special ritual concerning which Kloft, as we shall hear later, made some fascinating observations. The ants also protect the aphids against their enemies and carry them to safety if any disturbance threatens; sometimes they even shelter them beneath a protective canopy of earth or silk web. Finally, in temperate zones, the ants keep the eggs and sometimes the adult aphids and coccids in their nests and transport them on to the young shoots later, when the good weather returns. All these points deserve a more detailed examination.

Do ants kill their aphids?

Yes, replies Herzig, who has seen *Lasius* do so, but only when attack by a pest or predator has disturbed the aphid, causing it to make sudden, disorganized movements and drop some honeydew from its cornicles: then the ant will sacrifice it and carry it to the nest. It is also undeniable that in certain circumstances ants eat their coccid familiars, especially when the little insects are multiplying excessively.

How does an ant approach an aphid?

Although numerous myrmecologists have long been interested in the mutual behaviour of ants and aphids, nobody quite understood why, instead of devouring them, the workers treated the latter with so much of what might well be described as tenderness (if ants were not apparently the most unfeeling animals in the whole of creation). Heaven knows they have an appetite for anything that falls into their clutches. It was Kloft who first made an astounding observation in connection with an equally strange theory—and yet it could be true. According to him *the rear end of an aphid closely resembles the head of an ant*: still more so when, on the ant soliciting it for honeydew, its back legs are raised and frame its abdomen like antennae: this reaction can, in fact, be released simply by brushing the greenfly's abdomen with a paint brush. *The ant solicits food just as it would do face to face with one of its fellows*, and that stimulates the emission of a drop of honeydew. Moreover, *gorged ants have also been observed offering nourishment to the back of an aphid*! The resemblance is helped by the ant-kept aphid's usual lack of cornicles, which would spoil the likeness if they were present.

These are very important conclusions, but it must be remembered that they apply only to aphids and not to scale insects. What does an ant see when it inspects a colony of *Coccidae* but a series of flattened cones with no resemblance at all to the front of an ant? Nevertheless she solicits them with antennal pattings just as eagerly as she does the aphids and with comparable success. It must be added that she also pats the abdomens of individuals that are dead or too young to yield honeydew (Fig. 20).

This constant stimulation has a marked effect on the aphids, greatly increasing their secretion of honeydew and hence also their intake of sap. Herzig found, for example, that *Aphis sambuci* ejected 9·72 cubic millimetres of honeydew a day when milked by ants, but only 3·24 when unattended. He estimates that ants cause the secretion to augment by anything from 30 per cent to 300 per cent, according to the particular case in question. The *quantities taken* by the ants can easily be measured by observing trees on which the aphids are frequented by red

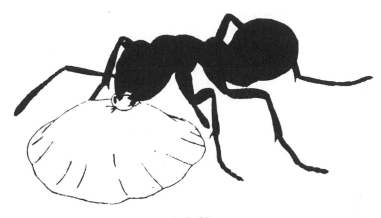

FIG. 20
An Argentine ant sucking up a droplet of honeydew excreted by a
scale insect (after Ghidiri-Pavan, 1957).

ants. It is soon established that on the trunk of a pine where
workers are climbing up and down, the ascending individual
is "thin" while the abdomen of the descending one is swollen
with honeydew. It then only remains to weigh the one and the
other and work out an average. In this way Œkland and
Zoebelcin have shown that an ant carries away an average of
5·5 mg. of sugary juice containing 10 per cent to 20 per cent of
dry residue. A colony of a hundred thousand individuals
sending out an average of twenty thousand foragers to lick the
aphids five times a day brings in a hundred grams of honeydew
daily, that is an annual fifty kilograms of liquid containing ten
kilograms of dry residue. Zœbelein, however, goes much
further, in pointing out that these figures apply only to the
relatively small colonies of *Formica rufa*. If we turn to *Formica
polyctena*, with its much larger populations, usually passing the
million mark, we get imports of *five hundred kilos of honeydew
per year, equivalent to a hundred kilos of dry matter*. The same author
has shown that *from a single pine-tree* the ants gathered two
hundred and fifty grams of honeydew daily: to express it
differently, each pine-tree yielded an annual four kilos of dry
matter to the ants. One would suppose that the tree would be
exhausted. We shall see that it is nothing of the kind, owing to
special characteristics of the aphids attractive to the red ant.

Other honeydew addicts

The strange thing is that *ants are not the only insects interested in the aphids' honeydew*. We will not count the many species that lick the honeydew scattered over the leaves by aphids who, as was mentioned earlier, project it to some distance when unattended by ants. The ant is not alone in discovering that, if properly asked, the aphid will give anything desired. Herzig has on several occasions seen flies succeed, using the same method as the ant, that is by patting the aphid's abdomen. The tricks of these flies (*Desmometopa, Leucopis, Tephritis, Rhagoletis, Fannia*, etc.) were well known to several early entomologists, but seemed difficult to understand. But it is not only flies that "milk" aphids; Coleoptera such as *Coccidotrophus* and *Emansibius* (which have incidentally been described as sub-social) do the same; as does also a Lycaenid butterfly (related to those blue butterflies that hover over the heather in summer) of the *Allotinus* species. Another myrmecophilous Lycaenid, *Gerydus boisduvali*, licks up the aphids' and coccids' honeydew while the ants are occupied with itself and its larvae, licking the special secreting hairs (trichomes) upon which they dote.

Are ants useful to aphids and scale insects?

It cannot be denied that they are so, and it would be tedious to quote here the numerous writers who have proved it.

First and foremost, they *actively protect them against their enemies*, and especially if these move quickly; for with ants it is almost always rapid movement that triggers off an attacking reaction—provided, of course, the said enemies are not equipped with some special defensive mechanism.

The wiles of the aphids' predators

There are, for example, the *Eublemma* caterpillars who conceal themselves in a cocoon made up of the scales of dead coccids very solidly laced together with silk web; thus disguised they can literally browse on the scale insects right in the middle of the ants who are milking them and seem in no way to resent the predators' presence. A Lycaenid butterfly, *Spalgis lemolea*, is also very fond of scales when it is at the larval stage. At this stage the caterpillar's flattened back looks so much like that of a coccid that it is difficult to tell them apart; it can devour them at its leisure, and the workers do nothing to stop it. Lastly, one of the Diptera, *Microdon*, who also has a taste for coccids, wastes no time in satisfying its appetite without let or hindrance. It simply becomes a guest of the ant *Solenopsis geminata saevissima*, which shelters a type of coccid within its nest. There the *Microdon* reaches its goal without the least concealment, and again the ants offer no resistance. Furthermore, there are parasites which the ants never touch, God knows why: a wasp known as *Encyrtus*, for instance, which lays its eggs at leisure in a scale highly esteemed by the ants (*Saissetia hemispherica*) while the workers not only allow it, but seem to draw aside when they meet one.

The guards who watch over the aphids are at once strict and indulgent in their habits. For example, the shepherds themselves never suck their beasts' honeydew. On the other hand, a parasite of the aphid *Myrmebosca mandibularis* sucks the honeydew without any objections from the guard, after which it goes to beg shamelessly for more from the workers, who give it what it asks! (Pontin, 1959).

Protective action by the worker ants

Nevertheless, in general, except in special cases like these, ants actively protect their aphids; and, as we shall see, in doing so they even set themselves against various attempts at biological control which would probably have every chance of success but for them. But *the efficacy of the protection depends on both the species of ant and the species of aphid or coccid*. There are some ants who can do little in this matter. Others, like the very aggressive Argentine ant, give their cattle one hundred per cent protection.

Protection against parasites and predators is not the end of the story, however. In most cases the aphids and coccids multiply much more quickly in the care of ants. The degree of success in aphid-raising, too, depends on the species of ant, and some are more effective than others. For example, Van den Goot found that after four-and-a-half months the average number of coccids was seventy on bushes with no ants, four hundred when *Dolichoderus* ants were tending them, but one thousand and fifty-seven under the care of *Anoplolepis*. There are also many cases where the scales and aphids disappear rapidly as soon as the ants cease to look after them. This is particularly true of certain species that are very harmful to our crops, such as a coffee pest of Kenya, which would not long remain a problem if the *Pheidole* ant did not lavish such disastrous care upon it. On the experimental level Flanders has beautifully demonstrated the part played by ants in connection with the orange groves of California, which are so attractive to coccids. He put a sticky band round the trunk, thus preventing the ants from climbing up. The coccids disappeared completely and quickly from the trees thus treated; because of a parasite that is highly effective when there are no ants to stop its activities.

Only Starý (1966) denies this theory in noting the presence of numerous parasitized aphids in aphid colonies, whether tended by ants or not; but we have already observed that the degree of protection depends on the species of ant and parasite.

The mechanics of protection

How does this beneficial action work? This is a question we do not always understand. But in some cases we are beginning to get an idea of it. There are some aphids and scales which secrete honeydew in such quantity that they are literally drowned in it: more especially the scales, which have no power of projecting the liquid away from themselves. After a while, particularly in the tropics, a host of yeasts and fungi grow in the sugary mass and poison these sap-suckers. Except, that is, when the ants are there to relieve the creatures of their excess honeydew. This is often seen, particularly in the case of coccids protected by *Œcophylla* ants. In fact, their solicitude, if I may call it that, goes further; if the scales seem to increase too rapidly (or for any other reason) the *Œcophylla* will not hesitate to consume the excess. This happens to such an extent that there is a constant ratio between the number of ants in a colony and the head of cattle they own.

In a significant number of cases the protection of the aphids and coccids is complete in so far as that their environment is perfectly controlled by the ants: they raise them either inside the nest, as we have seen, or in its immediate vicinity, making stable-like shelters for them out of earth or silk. Their protection against the elements is assured, but their safety from predators and parasites is not so complete since, as we have already heard, a certain number of these enemies succeed in penetrating the best-guarded fortresses.

Do ants carry their aphids about?

It was Lubbock who first raised this interesting question in 1882. According to him the ants known as *Lasius flavus* carry the aphid eggs into their nest in winter, and in the following spring put the newly hatched aphids back on the young shoots. On this point there is a whole series of somewhat contradictory observations. Some say they have seen the ants either carrying the aphid eggs to their nest or bringing out the young aphids in spring; others, observing the same species of aphids and ants, swear that there is nothing in it and that they have never witnessed anything of the kind. They point out that in most cases the ants concerned are subterranean species who raise root aphids, and these aphids, as we have seen, are kept in separate "stables" that must be considered more or less as extensions of the nest. How, then, can we ascertain that the ants carry the aphids about? Is it not more likely that the movements are part of the aphids' own behaviour?

Furthermore, when Herzig, by means of his sticky band, stopped the ants from climbing up the trees, he showed that just as many aphid fundatrices hatched there in spring as on the unbanded trees; *the winter eggs had been there all along*, but very difficult to see. It has also been said that the workers transport the aphids in spring on to the young plants, whereas they are observed, as often as not, *preventing them from getting out of the nest*. Finally, it is sometimes maintained that the young insects are carried back to the colony by their guardians when the weather is unfavourable, but even at −1°C Herzig found colonies of aphids outside, although the plants they were on were frequented by ants.

All the same, aphid eggs have undoubtedly been seen in *Lasius flavus* nests, where the workers were taking great care of them and licking them (there are no ant eggs in *Lasius flavus* nests in winter). But Muir made the strange observation that these eggs do not belong to the species usually associated with *Lasius*, namely *Forda formicaria* and *Tetraneura ulmi*. Then what

are these strange aphid eggs doing in the *Lasius* nests in winter?
It becomes apparent how doubtful and difficult to interpret
long-held theories can become in the eyes of more attentive
biologists.

Another undeniable fact is that ants carry away their aphids
if danger threatens. For instance, *Lasius brunneus,* which shelters
its herds attached to the bark of oak trees under paper-like
roofs, removes the insects to a dark place as soon as this covering
is removed. The *Dolichoderus* ants of Java have adopted a still

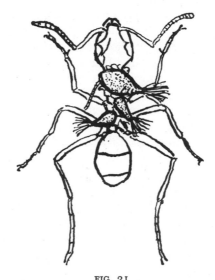

FIG. 21

Dolichoderus gibbifer carrying three scale insects (*Hippeococcus* sp.), one
2nd instar and two 1st instar (after Reyne, 1954).

more radical and amazing procedure: in case of disturbance they
tap their scale insects on the back with their antennae, at which
the coccids spontaneously detach themselves and climb on to
the back of the ants to be carried to safety.

Bünzli's phenomenon (1935)

There is also in my files a piece of observation that, magnifi-
cent though it is, has never met with the attention it deserves.
Bünzli has seen the virgin queens of the Javanese *Acropyga* ant
carrying in their mandibles on their nuptial flight the root

coccids with which their species is closely associated. The transported scales are always young and female; the ants never take males with them. Presumably, the colony-founding queen's first care after fertilization is to place her dear scale insect on a root, so as to be able to savour honeydew as soon as possible. This is very reminiscent of the behaviour of *Atta* queens who, when they depart to be fertilized, carry in the infrabuccal sacs a morsel of mycelium from their precious fungi.

Ants, aphids and agriculture

Generally speaking, ants are more harmful to agriculture than otherwise, because of their mania for raising aphids and scales.

But, before going any further, let us make an exception of the red ants. They, too, are addicted to aphids, and we have seen how zealously. But in this case *the aphids they tend do not harm the plants*. This is easier to understand if we go back to the two ways that aphids feed. The red ant aphids suck the phloem, that is to say the processed sap circulating in the sieve tubes; they inject no toxic saliva into the plant tissues, or almost none. Other aphids are parenchyma suckers; they insert their rostrum into the cells themselves and inject into the tissues an almost continuous stream of very toxic saliva that seriously upsets the plant's metabolism. By contrast the red ants' aphids do not inject the plant with virus diseases as is unfortunately so often the case. It is clear that the ants extract a considerable quantity of sap each year. But a German forester called Müller has calculated that all the sap taken by a nest of ants in one year equals a loss of wood growth worth only one mark, an insignificant amount.

On the other hand more than one biologist has conceived the idea of using the ants' predatory powers to rid crops of their pests. And up to the present only the red ants have given good results. There have been attempts to use other species against a Mirid attacking cacao in Java, against a damaging thrip in cacao and citrus in Brazil and against Sphingid caterpillars attacking coffee: people have even tried to set the savage *Prenolepis* ant against other ants, such as *Atta* which devastates orchards with its mania for collecting leaves to feed its fungi. But in every case these ants have paid far more attention

to cultivating their beloved aphids and scales than to the task
they were supposed to perform. Now, nearly all these creatures
are carriers of virus or fungus diseases which seriously damage or
kill their host plants.

So much was this the case that further attempts to use ants
in this way were abandoned: possibly too quickly in my opinion.
One might have tried giving the ants sugar syrup, as one does
to bees: would their taste for aphids and honeydews have remained so intense? Or could one not get rid of the aphids by
using a systemic insecticide, which is absorbed by the plant
and only kills the aphids and not the ants? Possibly one treatment would be enough and the ants without aphids and avid
for sugar might agree to police the orchard for the rest of the
season.

Could one not use Œcophylla?

There is, perhaps, a little hope with regard to Œcophylla.
As far back as the thirteenth century the Chinese were
putting nests of Œcophylla into citrus trees as the young fruit
formed, in order to protect it from pests. They had noticed,
well before us, that these ferocious weaver ants clear all before
them on the trees on which they live. They attack practically
everything, even the driver ants, which are, however, un-
conquerable: to do this they simply mass themselves on the
edges of the colony and seize all workers leaving it. The
soldiers do not seem to notice what the Œcophylla are doing
and never attack them, and the Œcophylla in return leave them
alone.

Obviously we now come to the question of the weaver ants'
herds, for they keep them as do the other ants of which I
have written, and are more eclectic than any: they are attracted
to practically any insect that secretes honeydew. Fortunately,
the harm done by these herds varies according to the particular
plant in question. For instance, they can cause considerable
damage to coffee and cacao but very little to mango and coconut. The ancient Chinese were able to avoid this basic difficulty
for the climate of southern China is too severe to allow these
ants and their guests to establish themselves; thus an excessive
increase of pests is prevented. In any case, according to Way,
the part played by the Œcophylla on coconut trees is highly

beneficial, and they drive off a pest called *Theraptus* which ruins the trees in certain areas. The only difficulty (nothing is ever simple in biology) is that certain other ants attack the *Œcophylla*; but these intruders can be controlled. Thus, always provided one first makes a careful study of their domestic animals, both *Formica polyctena* and *Œcophylla* can be included in the list of man's friends.

Ants and biological control

In any case, there is no doubt that ants often seriously interfere with attempts at biological control. For instance, a new parasite or predator whose efficacy has been proved and which has been bred up and released in mass, may be introduced into a new environment. Its success in one environment and set of conditions has been confirmed; and then suddenly, in new surroundings, its failure is almost complete. Ants have intervened. For example, the ladybird *Cryptolaemus montrouzieri* is very active against certain scale insects on citrus, and it has even been raised on a commercial basis. But nothing can be done if the Argentine ant is already there: the ant must be got rid of, or it will undo all the good done by the ladybirds. Unfortunately this ant is very difficult to destroy, because it forms polycalic colonies with many subdivisions that cannot all be reached at the same time. Also one finds scale insects that cause no damage, or very little, in one country where the ants are not interested in them, which become pests in another as soon as the ants exploit them, drive off their natural enemies and prevent them from performing their beneficial task.

CHAPTER IV

CITY LIFE

*The nest — Care of the brood — Making queens —
Division of labour*

Ants' nests

IN GENERAL ANTS' NESTS are not particularly remarkable as
regards materials or regularity in construction: it is quite
otherwise as regards size, for they are, apart from termite
nests, the largest constructions made by insects. The majority
of them are excavated in the soil or some workable construction
material; and it must be noted that *Formica fusca*, unlike
many burrowing insects which dig out the earth with their
mandibles, burrows by means of its forelegs and pushes out the
soil as does a dog. At this point we should note a superstition
to which biologists have clung for ages, and to which some may
still cling, like an oyster to its rock, that of "specialized organs".
My early scientific years were nursed on this theory: the burrowing feet of the mole cricket, like those of the mole itself, used
in the same way, were quoted, the predatory legs of the praying
mantis to grip its victims, etc. From which it was concluded
that animals had no need of tools for far-seeing Nature had
provided them in the very structure of their bodies. Very nice,
only it is wrong: the only really effective and universal tool
that Nature is at all concerned with is instinct. If the indispensable instinctive mechanisms have been established there
is nothing to worry about! Insects will burrow very well without digging feet and will seize their victims without predatory
arms. Moreover, the insects which move most earth, such as
Messor, Formica polyctena and *Atta*, are quite without special
digging organs and there is not a fiercer hunter than the red

ant, which has no trace of a predatory arm: numbers and instinct take the place of specialized organs.

But, apart from the red ants (see p. 34), we have no good modern studies of nest construction in ants. Yet this subject provides some interesting material, as Huber has shown. For instance, *Lasius niger* makes a mound of several storeys, consisting of chambers and galleries. Each storey is only about 10 mm. thick, with a floor and ceiling of about 1 mm. The ceiling is supported by little columns and walls. When these insects want to add another storey, they bring pellets of earth to above the topmost storey and build walls, columns and a ceiling. *Formica fusca* builds to a similar plan, but uses a different method. The ants start by piling up a mass of solid earth on the top storey, which they then burrow into in slices; later the solid earth remaining between the slices will be excavated and provided with a ceiling. As Sudd colourfully put it, "The *Lasius* nest is made of walls and that of *fusca* of holes. . . ."

The nests of the *rufa-polyctena* group (see p. 34), formed of a

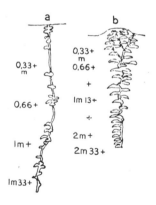

FIG. 22

Two kinds of vertical nest: (a) *Prenolepis imparis* (after Talbot, 1943); (b) *Pogonomyrmex baduis* (after Wran, 1938).

maze of underground chambers roofed with a pile of twiglets, display some strong adaptations to circumstances. The Germans export a number of red ant nests to the Italian Alps where there are not enough indigenous ones to protect the forests.

Now the German ants come from fir-tree plantations and have to build their new mounds with the Appenine pine and larch needles. At first they do not like the new material and go long distances to find the fir needles to which they are accustomed. Gradually, they start collecting a few pine needles, but are content to leave them at the base of the mound; later on they incorporate the new material in the mass. Pavan asks whether these ants have really adapted themselves to the use of a new material, or whether the young ants born after the moving of the nest do not straight away use pine needles simply because they have never seen any others. It is a small matter, but not without importance, above all if our first hypothesis is correct: we can conclude that if the *ants' memory is capable of registering the characteristics of their usual materials, their behaviour is not as fixed* as one might imagine and is capable of change.

Other ants use dead or living vegetable matter. The former is, for the most part, colonized by a wide variety of species, such as the red ants and *Leptothorax*. Inside the red ants' mound there is always a *more or less* decayed tree stump or root. The ants make use of all the holes in it, as well as making more for themselves over the years. In the end the wood is reduced to a lace-like network of dividing and interconnecting galleries. The smallest offshoots are inhabited by the little *Leptothorax*. *Camponotus* use the trunks of dying or dead trees in a very different way. Avoiding the toughest bits, the ants burrow through the soft part of a tree's concentric rings. They do this so accurately that they finish up with a series of concentric shafts, *which can reach a height of ten metres*, without any horizontal dividing walls. The larvae are attached to the walls here and there by special hairs.

Living plants can be inhabited by *Colobopsis*, whose larger workers, *majores*, have a very specialized and enormous, flattened head. These big workers or soldiers stay inside the nest where their heads exactly block the entrance gallery. Who could fail to be amazed at such marvellous adaptation? No less remarkable is the fact that the entrances of the red ants' nests are just as well guarded, despite the fact that their workers have no special morphological adaptations of the head or any other part. Here is another example where it is not a question of special adaptation, but of the number of workers involved and, above all, of their instinct. Nevertheless, a great deal of fantasy has been expended on this subject: there was a theory, for instance, that a special tapping code is applied to the head of the gate-keeper, who opens to the correct signal. It is not impossible; but other species, whose *major* workers also guard the entrance, but are without (and this is important) morphological adaptation of the *Colobopsis* kind, leave their antennae outside, which allows them easily to recognize the smell of their compatriots.

A still more curious case is that of *myrmecophilous plants*, which have made-to-measure cavities or ones that the ants can adapt with but little effort, and which are always associated with the same species of ant. Many thick stalks with friable pith are thus colonized not only by ants but by many other species of insects. Moreover, the ant plants generally seem to flourish, perhaps because the workers free them of pests. Certain trees, such as the *Cecropias* of South America, are often infested from top to bottom by *Azteca muelleri* who have replaced the pith and seem to feed on the small swellings, the Müller corpuscles, which are to be found at the base of the leaves. Another species, *Pseudomyrmas*, likes to install itself in the South American locust tree's huge thorns, which swell up like oak apples, though the ants do not appear to be responsible for that phenomenon. An epiphytic Javanese plant, *Myrmecodia*, has a very distended bulb, which is filled with a large number of natural cavities. This is the favourite home of *Iridomyrmex myrmecodiae* which even deposits its excrement inside the plant, thus providing it with additional nitrogen. Certain species go so far as to colonize those entomophagous plants, such as the American *Sarracenia*, which have a kind of gourd, that traps and digests any insect venturing inside it. It seems that ants who risk entering these plants escape death only if they choose a fairly old or dry gourd, where they can thrive, while hundreds of their fellows perish in the neighbouring green gourds.

A few species make their nests out of a paper-like substance. *Azteca*, for whom the protection of myrmophyllic plants is not enough, builds a thin paper nest in the hollow where the pith used to be. In France the best paper-making ant is *Dendrolasius fuligiosus*, the black wood-ant. Its highly complex contructions consist of finely-shredded wood fibres mixed with

FIG. 23

Section of a *Myrmecodia* plant from the Bismarck Archipelago,
showing cavities occupied by ants in the pseudobulb.

saliva. No one to date has gone to the trouble of observing how
Dendrolasius sets about this task.

Next, I should like to mention the *Crematogaster* ants of
Africa, whose constructions are among the most noteworthy in
the ant-world. They are made out of a mass of paper cells and
can be as much as 1·75 metres high and 1·30 metres across. In
some species the paper is mixed with earth.

Silk nests

The nests of *Œcophylla* or weaver ants can be found in an area stretching from Queensland to East Africa. They sometimes form quite vast sheets of layered, stuck-down leaves. There is also *Polyrachis*, which makes at least partial use of silk to reinforce its nests, but not with the same degree of care that *Œcophylla* uses.

I have dealt elsewhere with the manner in which they do this. It is particularly strange because it is the larvae that have silk glands, and not the workers. They use the larvae in somewhat the manner of a loom shuttle: seizing one in their jaws they systematically touch the edges of the leaves in question with the end of the silk canal, leaving a sticky thread at each point until a complete tissue is built up. This is important, for what they are doing is *neither more nor less than using a tool*, a highly unusual thing in the animal world; important also because it was once said that the use of tools was undoubted proof of intelligence. We have grown more cautious, especially since it has been realized that the word intelligence is hard to define and covers a whole series of probably very heterogeneous functions. Moreover, we must distinguish between the isolated chance use (probably innate in the species) of a tool that is employed by all without any further development, and the *systematic use of a number of tools that are evolving all the time*: only the latter can be considered indubitable proof of mechanical intelligence. With animals only the first has ever been observed, never the second. Their tools are very simple, always used in the same manner, and very rare at that. Apart from the weaver ants' shuttle there is the pebble with which some Hymenoptera stop up the entrances to their burrows; the straw used by chimpanzees to extract termites from their holes; the thorn that the Galapagos finch employs to scratch larvae out of cracks in wood; the little stone grasped by the sea-otter to break open the shell-fish that are its food; the wad that gardener birds find useful for painting their breasts—and that is about all that is known of this kind of behaviour. It would seem that evolution has not tended towards tool-making, with the obvious exception of mankind's.

Let us go back to *Œcophylla*. Before the weaver begins her

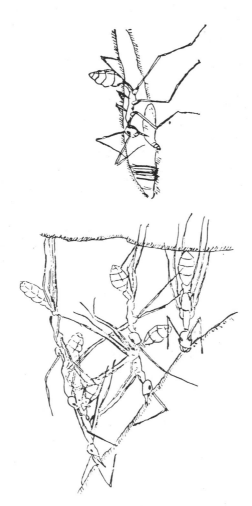

FIG. 24

Left: Œcophylla drawing together the edges of two leaves; *right:* a worker uses a larva as a shuttle to apply silk threads (after Ledoux, 1949).

task, teams of workers, stretching themselves out enormously and climbing one on the other, succeed in reaching the edge of a leaf at some distance from the one to which they are attached. Immediately long chains of weavers start to draw the two together. Here we have behaviour typical of ants and social insects. It would be naïve indeed for a layman to claim that this work involves intelligence on the part of the ant. In fact, as soon as the colony is uncovered the workers seem to be seized with a compulsion to pull on absolutely anything. We have already pointed out this unco-ordinated activity in the case of the transport of victims by red ants, for instance. Nevertheless, it must also be emphasized that the task is eventually carried out with speed and skill in spite of the absurd disorganization of some individual actions. For instance, if the nest is made of a single folded leaf, it is always curled over so as to cover the brood and not in the opposite direction. So the weaving is random only in appearance; thus the "mean distribution" is not as chance will have it but directed to a definite end. Moreover, *Œcophylla* can use any kind of leaf, from the narrowest to the widest.

Care of the brood

It is the nature of all social insects to care for their young in common. To be accurate, ants are not the only ones in the insect world who show maternal instincts. In the case of earwigs or mole crickets, for instance, it is fascinating to observe the mothers watching over their eggs for days on end. Lhoste has shown that, as far as earwigs are concerned, the eggs cannot hatch without the presence of the mother, and, if she is taken away from them, they soon get covered with mould. The mother's saliva must have some antibiotic action, since similar substances are often secreted by insects. However, neither earwigs nor crickets take any notice of their young larvae, although they also never behave aggressively towards them.

In the case of social insects, however, one cannot generally describe the care they bestow on the brood as *maternal*. It is true that in certain species of ants the female or queen looks after her eggs herself during the foundation phase, when she is alone; but this is not usual and often the female has to be assisted by the workers of her colony, or even those of an alien colony. . .

As to the workers they are, of course, not the mothers of the larvae, but at most their sisters or half-sisters. It is not impossible, as we shall see, for the workers of some species of ants to lay eggs capable of development, but these are incorporated into the common brood and not tended by their mothers.

However, contrary to what is generally supposed, care of the brood is not universal among ants. There are some exceptions among the most primitive species. *Myrmecia*, for example, does not lick the eggs or look after them in mass. Likewise, although as a rule larvae regurgitate drops of moisture that are avidly sought by the workers, this does not occur among the archaic Ponerines. Some nymphs have no cocoons, but with those that do, such as the red ants of Europe, the workers help the nymphs to tear the thick silk that prevents them from hatching. We do not know what signal warns the nurses that the time has come to open the cocoon. Raignier has observed that, if for some reason the nymph inside is dead, its cocoon is still torn open in the same way on the correct date for eclosion. But in the primitive family of the Ponerines the nymphs have to manage by themselves.

An increasing development of social interactions has also been noted, not only as regards adult-larva relations, but also of those between adults. The trophallaxis or exchange of food, so common among higher ant species, is very little developed in the case of Ponerines. Nor do we find with them the transport of one adult by another, that activity so usual among the higher ants, although still mysterious to us.

The influence of the young on the adult

On the other hand, and this happens with all animals, right up to the mammals, *the care of the young is never a one-sided affair*, in the sense that the young exercise a very special effect on the adults: the egg or chick modifies the adult bird's behaviour profoundly; the young one disturbs the hormonal balance of the mammal that feeds it. In the same way the brood is a central factor influencing the behaviour in an ant nest. We will leave aside the part played by the brood among hunting ants, where it regulates the entire behaviour of the colony (page 76), but even with other species social cohesion is barely maintained without the presence of the brood. It is, above all, the movements and food requirements of the larvae that appear indispensable to the workers—which inevitably reminds us of the driver and army ants.

How the workers make queens

Queens and workers. We find the same phenomenon with ants as with all social insects, namely one or several reproductive females, wrongly called queens (for they do not rule anything), and a sometimes enormous multitude of workers, who are in fact imperfectly developed females. The two sexes are found in the queen's *entourage* only among termites: here there are male and female workers, both incapable of reproducing their kind, and whose reproductive organs are so shrunken as to be almost imperceptible.

There is nothing very special to be said about the ovary of an ant queen. It is made up of a number, greater or smaller according to species, of *ovarioles*, little tubes where the eggs ripen progressively from front to back.

During the nuptial flight a sometimes enormous quantity of sperm is stored up by the queen, who mates a great number of times. The sperm will remain alive during the many years of the queen's laying period, which is sometimes as long as ten years, and continue to fertilize countless millions of eggs.

If we take the red ant, by far the best studied species, as our example, we note a curious difference between *monogynous and polygynous* communities. The queens of the former have several hundred ovarioles, enabling them to lay nearly three hundred eggs a day during the five or six months between the end of March and September. The queens of polygynous communities, however, have only 90 to 170 ovarioles and lay only about ten eggs a day. It is true that more than 5,000 queens may be found in a single colony, which represents the formidable increase of nearly 9 million ants per season. Since the workers live three years it is clear that if all these eggs actually hatched, our woods would be entirely covered with a seething carpet of ants; however, a considerable number go to nourish the first-stage larvae. Nevertheless, in experimental formicaria of *polyctena* where the eggs can be easily seen, I have often observed that there were enough to fill a soup spoon. And I mean *eggs*, and not pupae, which are so often called eggs.

In worker ants the atrophied ovary has broadly speaking the same structure as in the queen, but consists of only two to six ovarioles, the *major* workers having more than the *minor*. It must be added that when no queen is present the workers of monogynous and oligogynous species can lay unfertilized eggs that produce males, whereas with polygynous species that is impossible, or at least very rare.

Not all workers are capable of laying eggs; for that their ovaries must have reached a certain stage. At the adult's eclosion they are hardly visible, and it is only between the 11th and 27th day that they attain their maximum development with the presence of eggs. Normally, however, these are not laid but reabsorbed so that all the nutritive substance goes back into the worker's body. Nothing remains in their place but a yellow spot in the ovariole. The process of reabsorption is very variable from worker to worker and may take days or months.

All this is very similar, even to the number of days, to what takes place in bees. As with bees, the fertilized eggs produce females (queens and workers) and the non-fertilized eggs produce males: that is *Dzierzon's law*, well known with reference to bees, where Dzierzon first observed it. The phenomenon is, indeed, more complex in the hive, where the queen bee possesses the extraordinary quality of being able to lay at will worker eggs in the small wax cells, or male eggs in the larger ones. This process involves a complicated organ called a sperm pump as well as a special awareness of the cell's width, which can only be achieved with the aid of sensory organs of the integument. Things cannot be quite the same in the ant world, where there are no cells, but nevertheless, a queen can lay male eggs at certain well-defined times, such as the beginning of the year, in the case of red ants. Hence she must be able to control the insemination of her eggs to a certain extent.

In any case, Dzierzon's law has been verified in regard to thirty species of ants, with the possible exception of *Lasius*. With them it is just possible, though rare, for eggs laid by workers to produce females, although these eggs are unfertilized. Otto maintains that the same thing can occur among *polyctena*, even fairly frequently. It is said that the unfertilized eggs of a South African species of bee, *Apismellifera capensis*, can also, on occasions, produce females.

However, in the sex-determination of red ants a special factor intervenes, relating to conditions in the nest. *Large colonies, well heated by the sun, produce predominantly females; smaller ones in the shade produce more males.* In fact, the former can be made to produce males by providing artificial shade. Below 20°C, the queen lays unfertilized eggs, above it, fertilized ones. The lower the temperature sinks (down to 15°), the greater is the percentage of unfertilized eggs. All this, however, is only true just after the winter. In summer laying goes on between 8° and 30°, and only workers are born.

Queen-producing and worker-producing eggs

But we may well ask ourselves, as with bees, why some of these apparently identical eggs produce queens, while others give rise to workers. Is it due to some intrinsic property of the egg? In other words, is it destined from the time it is laid to produce a queen? Or, on the other hand, is it a matter, as with the bees, of an ordinary egg, which will develop into a larva fated to be a worker unless it is given special nourishment (royal jelly in the case of bees)?

With ants there is no simple answer to that question. Here we find a wider range of possibilities and more complicated biological and social mechanisms; and more than one example of this peculiar characteristic of ants will be cited. First of all, the summer and winter eggs are different. The latter are surrounded by big nutritive cells which degenerate only very gradually, and the egg, 0·73 mm. in length, weighs 0·059 mg. The summer eggs, on the contrary, have few nutritive cells, and these degenerate rapidly, so that their length seldom reaches 0·62 mm., or their weight 0·045 mg. Also to be seen in the egg is a singular differentiation called "polar plasma" (the *Polplasma* of the Germans). Situated at the back, in the winter egg it forms a mass measuring 6,500 thousandths of a cubic millimetre, and is very rich in ribonucleic acid, whereas in the summer egg it measures only 800. Thus the winter egg would seem predetermined to produce queens; yet it is not a question of "blastogenic" determination, as geneticians usually suppose, for it does not depend on any property of the chromosomes, but *only on the way the queen is fed.* Moreover, this determination is not irreversible. Nutritive factors can play their part, it seems, after as well as before laying: if the early spring food is poor and inadequate, the winter eggs will yield only workers. As to the summer eggs, they may appear at a time of plenty, when the larvae are well nourished, in which case they will develop into forms intermediate between queen and worker.

Special food for the queens

But the explanation becomes more involved. Gösswald and Kloft have shown that there is a difference not only of quantity but also of quality between the food of queen larvae and worker larvae. The worker larvae are mostly given regurgitated food from the crop, whereas the nurses feed the sexed larvae with the contents of their big labial glands which stretch back right into the thorax. Gösswald and Kloft established this with the help of radio-isotopes, using the labelled-isotope technique. I have seen this process at our nuclear research station, and it is very interesting. The radioactive substances in the different organs affect a photographic plate in proportion to the concentration of the radio-isotope present. When they have been ingested by a rat it only has to be frozen at $-40°C$, and then divided longitudinally with a saw, to obtain a perfectly clean, exact section of the whole organism, with every organ in place. It then only remains to apply a photographic plate to it to obtain an autoradiograph. The same can be done with an insect.

In these conditions Gösswald and Kloft observed that the food was first stored in the ant's crop, after which it gradually passed into the intestine before becoming part of the general metabolism. It is especially concentrated in the labial glands. If, then, a hungry worker or larva is presented to a worker who has, as yet, deposited the radio-isotope only in her crop, she will distribute its contents to the other, who will soon share her radioactivity. If, on the other hand, she is placed in contact with larval or adult queens and feeds them in the same way, they will not become radioactive, because they will have received food from the labial glands and not from her crop.

These labial glands develop particularly (as do also the cephalic glands, but to a lesser extent) in spring, when the ants first ascend to the surface of the nest to bask in the sun. This is the time that queen-raising begins. When the tempera-

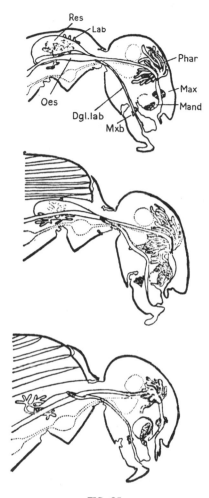

FIG. 25

Top to bottom, head of *F. rufa*, worker, queen and male. Mand.
Mandibular gland; Max. Maxillary glands; Phar. Pharyngeal
gland; Lab. Labial gland and sac (Res.); Dgl. lab. Labial gland
canal; Mxb. Gland at base of maxillae; Oes. Oesophagus (after
Bausenwein).

ture rises above 20°C, however, activity of the labial gland decreases rapidly within a few weeks; consequently the workers are no longer able to breed queens. All the same, with certain red ants who have a second period of raising sexed individuals during the summer, the labial glands regain their activity (Fig. 25).

Competition for food

The larvae destined to be sexuals demand so much labial-gland secretion that it is impossible to bring them up in a community of less than thirty individuals, and then only if the queen is absent. When the queen is there the demand for food rises sharply: with *Formica nigricans* at least 600 workers are needed to maintain the royal brood; with *rufa* or *polyctena*, when a single queen is present, it takes 3,500. One wonders how they manage to raise numerous sexed individuals every year in *polyctena* colonies with thousands of queens.

Gösswald thinks it is made possible by the retreat of the queens far from the rest of the community into the coolest parts of the nest after they have sunned themselves like all the others for several days. And, indeed, if cold zones are created in experimental formicaria, the queens come to hide themselves there and the production of sexed larvae begins at once. But what happens, exactly? There are two hypotheses: either an alimentary independence from the rest of the colony is established between the queens and the workers who accompany them (for they must still be nourished as they have not stopped laying), which could doubtless be verified by distributing radio-isotopes to the queens' group—if the supposition is true, these should not later be traced among the rest of the community. Or else the removal of the queens weakens the effect of their odour on the workers, an odour which is supposed to have the property of hindering the procreation of new sexuals. The influence of the queen's odour on bees is well known, and several of its components have even been chemically isolated. With ants, however, the queen does not appear to exercise as much influence. For one thing, she does not attract the workers to her as strongly as does the queen bee, and especially not among polygynous species such as *Formica polyctena*. In short, the influence of the ant queen's odour has not yet been clearly demonstrated.

It only remains to add that, as with bees, the orientation of larvae towards a "royal future" can only be effected up to a certain moment. From seventy-two hours after hatching (at 27°C) it becomes too late. Males develop in conditions similar to those for females, but from unfertilized winter eggs.

How long does it take to make an ant?

Obviously it depends on the species. In the case of red ants it takes thirteen days at 23°–27°C for the eggs to hatch; the larvae pupate eight days after hatching, at 30°C, and the imagos emerge as perfect insects fourteen to sixteen days later, that is to say thirty-seven days in all.

What I have written above concerns red ants, but it also seems to apply pretty generally. With *Myrmica*, which has been particularly well studied by Brian, the queen must have some inhibitory action on the production of other sexuals, but this question still remains as uncertain as it does with the red ants. All one can say is that the presence of a queen reduces the size of eggs in the workers' ovaries and affects the growth of larvae: the smell alone of a dead queen does not have this effect, unless dead queens are presented to the workers at regular and short intervals. As to separating the queen from the worker by a metal mesh partition, with the object of studying the mechanism of her behaviour, this causes so much disturbance that no conclusions can be drawn. On the other hand, in long-orphaned nests new "laying" workers start to appear; they are very aggressive and firmly oppose the introduction of a new queen; except if one gives each one three larvae to care for: then they will lay no eggs and be friendly towards a new queen.

As regards their possible development, it seems that larvae can evolve towards becoming either "queens" or "workers" up to a certain point in the third instar, which Brian relates with precision to the progressive deplacement of the larval brain moving little by little to the thoracic segments, as can easily be seen by examining the larval body by transparency. All this depends on a series of factors: for instance time of year. In spring a considerable proportion of the larvae, some of whom, moreover, have had their development stopped during the whole winter (diapause), have a good chance of becoming queens. The workers also affect the phenomenon and their metabolism varies according to the season of the year: for example, spring workers induce continuous larval development whilst autumn workers bring about a diapause in nearly all individuals.

But the phenomena do not develop in the same way in the case of weaver ants. As Ledoux has pointed out, the queen and workers both lay eggs; the worker eggs do not follow Dzierzon's law found among bees, according to which unfertilized eggs give rise to males and fertilized ones to females; that is to say the worker eggs, though not fertilized, nevertheless give rise to females, queens or workers. As to the queen, it is not she who lays the eggs giving future queens; she only produces workers. Males come from chambers situated at some distance from the main nest, which the queen never visits, and arise from eggs laid by workers. We might thus conclude that the queen's role is the *inhibition of male production*. Moreover, the workers are of two sizes in this species, *minores* restricted to nursing the brood and *majores* to foraging. This seems also to be a case of regulation by food supply, because an isolated queen without nurse ants gives rise only to *minor* workers; whilst if she has nurses from the beginning, she produces *majores* and *minores* at the same time.

Division of labour

This is a fundamental feature of insect societies but it is still the subject of much discussion.

Let us summarize what happens with bees. Some years ago Rösch put out a very clear-cut theory, according to which the division of labour depended on the age of the workers: they were first of all nurses, then wax-makers, cleaners, honeygatherers, and finally guards. Then Lindauer proved that these categories were not so exactly defined and that some workers could skip one or more stages: a bee might well become a honey-gatherer without ever having been a wax-maker. In reality it is the needs of the whole hive which regulate the numbers of the different social categories, by methods we do not know very much about. One of them undoubtedly is the continual exchange of food from mouth to mouth: this enables all the workers at any one moment to know the contents, both in kind and quantity, of the "social stomach".

All research concerned with the division of labour depends on marking individuals, which thus become recognizable. This is not difficult with bees, and von Frisch established the principles by means of coloured spots on the head, thorax and abdomen respectively, representing units, tens and hundreds. By combining colours several hundreds of workers could be marked.

It is much more difficult with red ants; they are much smaller in the first place. However, Otto succeeded in sticking numbers and letters on their abdomens without, apparently, inconveniencing them much. Other workers have even managed to put little metallic rings round the petioles!

Internal and external work

The results are fairly complicated. Three categories of worker are found:

1. *Strictly specialized workers,* always doing the same work for months at a time (for instance, milking aphids).
2. *Unstable workers,* continually changing their tasks.
3. *A large number of ants with only a limited amount of specialization.*

All ants have one point in common, which is that they remain in the nest for the first period of their life and only forage outside during the second and last period.

The time of transition from one to the other can be very variable. Some ants only stay in the nest for three or four weeks: others have not left it at the end of a year. So that at the start of the eclosion of the pupae (beginning of June) there are workers in attendance who will become outside workers that same year, and others who will not go outside until the second summer of their lives.

Outside the nest, however, specialization sometimes appears so loosely defined that writers such as Œkland and Kül have maintained that it is practically non-existent. Their research procedure was to mark every worker along a red ant route with a different colour according to the kind of task at which they found it: transport of prey or twigs, tending of aphids, etc. On the following day they were unable to find any correlation between the colours and the tasks of the day before. On the other hand, Otto resumed this investigation but confined it to limited areas and terminal points where he saw ants at work: a colony of aphids on a branch being fed by workers, for instance, or a very narrow hunting territory provided with plenty of caterpillars or such. The workers having been marked, it was established that they showed remarkable constancy in their tasks, often for more than a month together; some frequented the same group of aphids, situated 40 metres from the nest, for forty-three days.

Enormous differences in individual reaction

At this point in his researches Otto uncovered a feature of ant behaviour to which, for my part, I attach the greatest importance. It would not take much for me to maintain that it is the basis of ant social organization: it is the enormous difference in individual reactions. For example, if Otto offered living prey or a cocoon of its own species to a milker of aphids, it paid no attention; and yet, Heaven knows, a cocoon can be a powerful stimulus to a worker. This heterogeneity of reaction is demonstrated still more clearly, as Otto proved, when the ants are attacking their quarry or imbibing syrup from a saucer. For instance, he offered a caterpillar to 160 workers. Only 25 per cent fought it to the death; 33 per cent fought, but not to the bitter end; 25 per cent did not even attack it, while 17 per cent actually retreated as if afraid. Out of 455 workers who were sucking up syrup, 38 per cent left it well before their crops were full. And out of 1,527 ants to whom Otto presented a cocoon, only 25 transported it, even when it belonged to their own nest. Note, in passing, that when the cocoons came from a foreign colony only ten workers took an interest in them. Some writers maintain that the cocoons (and also the newly hatched young) are accepted by all colonies without distinction because they have as yet no social smell. Surely this proves the contrary, for the workers on this occasion seemed well able to tell one from the other.

I stress this fundamental heterogeneity because, at the end of this book, I intend to draw some conclusions from it that may well surprise you. Otto certainly recognized its importance. He tells how he once observed two ants, who had both just hatched, side by side in the same colony. He offered them a very powerful stimulus, a hungry larva: only one of them took charge of it; the other paid no attention at all.

Age, then, is not the root of the matter; yet we have known for a long time that with bees the specialized task depends on age, and that, for example, it is always the young workers who have to feed the larvae.

The ovaries, the feeding glands and specialization

All the same, we must remember one definite point in this rather fluid picture of the division of labour: there is some degree of age effect, both inside and outside the nest. Size has little influence on the type of task, although among red ants the *minor* workers seldom join in the transport of prey and prefer to look after the larvae; the smallest ones (*minima*) never leave the nest. Yet with *Formica sanguinea*, studied by Dobrzanska, the opposite rule prevails: the *majores* occupy themselves with the larvae while the *minores* go out hunting.

However, Otto has found one clearly marked though unexpected feature that is in exact correlation with the type of work done inside or outside the nest: that is the development of the ovary and certain glands. We have already seen that the ovarioles follow a particular cycle: they begin by developing rapidly and producing eggs only to reabsorb them later. But the duration of this cycle and its different phases is not the same with every individual. It is usual for ants on internal duty to have large ovarioles containing eggs, and they do not leave the nest until the process of reabsorption is complete. If this degenerative process is artificially accelerated, the moment of the worker's leaving for outside work is advanced by the same amount of time. In natural conditions the degeneration begins after several weeks or possibly not till the end of a year. This relationship between ovaries and behaviour is also found among bees; with them, too, the ovaries of the honey-gatherers are completely regressed, except in the case of the egg-laying workers. In the absence of a queen the workers' ovaries become greatly enlarged, and they begin to lay eggs, unfertilized and therefore bound to produce males. This does not stop them from going out to seek for honey. In a normal hive, however, the workers with big ovaries are the youngest, and they stay in the nest. The maxillary glands, which are homologous to the pharyngeal feeding glands of bees, have almost entirely regressed in ants doing outside duty. The same applies to the large salivary glands, which seem to be fully functional only during interior duty. The other glands, mandibular and post-pharyngeal, do not change. Evidently all this must be connected with individual variations in hormone functioning.

During recent years the science of insect hormones has made giant strides on which I have no space to enlarge now. Anyway I recall that a group of very specialized organs, the *corpora allata*, situated behind the brain, play a highly important part in hormone regulation. These organs seem to be in an identical condition in all the red ant workers, but this impression may be due to a lack of precision in our means of investigation.

The different activity thresholds

In short, the idea of division of labour is illusory: certainly there is a relatively small number of individuals dedicated to one particular duty, but with most of the others it is simply a matter of one worker manifesting a very different threshold from her neighbour when faced with a particular task. Otto takes as an example the situation where the top of the nest has been damaged by an enemy: naturally the builders, whose individual threshold for handling twigs is very low, begin repair work at once; but the shock, and probably the ensuing loss of heat as well, drive a crowd of workers to help in the reconstruction, who would normally have remained indifferent to all stimuli connected with twigs for the surface. This brings us to a conclusion which leads on to much else, as we shall see later: for an ant community to function properly it *must be sufficiently numerous* for there to be a reasonable chance that, when necessary work arises, the threshold of an adequate number of workers will be low enough for it to be accomplished. And that postulates an unexpected corollary: the *ant colonies that show the most varied psychology will be the most numerous, and vice versa.* This is, in fact, what we observe in nature, where agriculture has arisen only among the *Atta* with their enormous nests, and stock-raising is practised only by the red ants, who have certainly the most highly populated colonies in all antdom. There is no denying this variability of reaction.*

---

* Although the division of labour is fairly flexible among red ants, it is not so with all ants. This is well illustrated by an old observation of Lubbock's. He kept 200 *Formica fusca* workers in an artificial nest for nearly a year, and noticed that

Variability of disposition (polyethism)

In a community of ants or of any other social insects it is difficult to understand *the part played by external stimuli* and *by the reaction of each individual.* Sudd gives several examples so obvious that no one has really thought about studying them before. For instance, if the brood is attractive to ants, of which there is no doubt, how is it that some of them leave the nest to go foraging, notwithstanding? And when hungry ants leave in quest of food, why does not the whole colony go with them? When ants construct a dome of twigs why do the builders constitute such a small percentage of the population, while the majority take no part, etc.? The whole secret is that in any one colony there are probably not two individuals capable of reacting at the same time in the same way to the same stimulus. Some stimuli, indeed, do not seem to be even noticed by more than a small minority of workers.

That being so, we can distinguish with Sudd first of all a variability of disposition by caste. With *Pheidole, Messor* and many other genera we find *minor* and *major* workers (the latter wrongly called "soldiers"). We have seen that these soldiers are no more aggressive than the others; it is not their special duty to defend the nest. Moreover, very often, the difference of behaviour between workers of varying size is merely a question of degree and is not easy to define. For example, often, although there are exceptions, the *major* workers do not go foraging, and small colonies produce only *minores.* With other species the smallest workers, called *minima*, never leave the nest, where they look after the brood. It is the same with bumble-bees. When a nest of these insects is opened up it is astonishing to find there tiny bumble-bees, at least five times smaller than those outside, who never emerge from the nest. On the other hand, the *majores* are often more versatile in their duties: Gœtsch tells of some *Messor* who changed occupations nineteen times in ten days, while the *minores* only changed twice. This instability is also found among *Formica polyctena* and *Pheidole.*

never more than nine of them, always the same, went to forage in the outside world.

Finally, it should be stressed that this very flexible division of labour must be set against a *faithfulness to locality*, which does not limit itself, for example, to one tree, but to one part of one branch of the tree. And this constancy continues from one year to the next.

Variation of disposition (polyethism) according to age is well known among bees, where the young work as nurses, and the old as honey-gatherers and guards. It is much the same with ants: the young workers stay in the nest looking after the queen and brood, and never forage; they are, moreover, very timid and flee at the slightest unaccustomed stimulus. The older workers, on the contrary, are used to external dangers and much more prone to examine objects found in their way. Their usual reaction to aggression is attack. After all, social insects are the only ones to stand up to man. But there is one fascinating though unfortunately little studied phenomenon: *certain features of behaviour manifest themselves earlier when the young are in the company of the old*, to such an extent that past authors supposed the younger workers were being taught by their seniors. That is too good to be true; and whilst I am disposed to believe that ants are capable of anything, I would not believe that without good proof. In any case the presence of the old leads to the development of ovaries in the young: we know (see page 123) that there is an ovarian development cycle according to age in ants as in bees; as certain changes in behaviour are correlated with the condition of the ovaries, it is probable that old ants do not teach the young to imitate them, but simply hasten their physiological development, which brings about a quicker change towards mature behaviour (Otto 1958).

Here, too, polyethism according to age is not rigid, and it is not age alone that regulates the work of the young ants inside the nest. Note also that there is no permanent court surrounding the queen; as with bees, the workers that attend her change frequently. We know very little of what turns an ant into a specialist at one particular task. It may be an innate disposition, or it may be that at a critical point in its early life it had to take on some special work that henceforth became fixed in its behaviour. Outside the nest, too, some workers are much more inclined to hunt than to collect honeydew.

Variation in the response to a stimulus, which is the basis of polyethism, is easy to observe when ants are bred under experimental conditions, When a *Myrmica scabrinodis* nest is struck, for example, the dark ants, which are the oldest, are the first to scurry here and there in excitement, while the paler, younger ones are the last. Wallis demonstrated that with

Formica fusca a blow on the nest caused only a small number of workers—always the same ones—to emerge, and that they were also the first to go in search of food. Others would not come out unless the disturbance was much more severe, but they would willingly accept food from the foragers. A third group not only refused to come out, but also accepted food only with the greatest reluctance. This last group connects up with the numerous workers who do nothing except perhaps store the food (see page 149). It is true that if the community is hungry the proportion of active workers greatly increases while, on the other hand, an abundance of food reduces their activity. Thus it appears that, as with bees, it is the needs of the community that regulate the proportion of workers occupied in any one task. Young *Formica polyctena* workers, for example, begin to forage sooner if all the groups are composed exclusively of their contemporaries. With *Myrmica scabrinodis*, whose colour facilitates an exact estimate of age, the young (light-coloured) ants pay more attention to the brood if older workers are present.

This brings us to the still unanswered question of *initiative and imitation*. Of course I do not speak of conscious initiative in the human sense of the word, but once more, as always, of differences in reaction. It would appear, for instance, that when a worker begins to carry the pupae about she is "drawing the attention" of her colleagues, who will lose no time in "imitating" her. If I put those terms in inverted commas, it is because I am not very sure they are the right words to use here. Though, indeed, a thought-provoking observation was once carried out but never, alas, repeated . . . the experiment of Chen.

Chen's experiment

In 1937 this Chinese biologist was studying *Camponotus*; he very soon noticed that when two of them were digging the earth together each worked much more quickly than when alone at the same task. Of the two ants, one was often more active than the other, and remained so during the few days that the investigation lasted. What is more, a "lazy" ant will dig harder when put with a more active companion. And, much more recently, Sakagami noted that *Formica fusca* and *sanguinea* dig faster when they are in large groups than when they are in small.

CHAPTER V

THE VICES AND ENEMIES OF THE CITY

Parasites and predators — Addicts — Slavery

Parasitic guests

ONE OF THE most puzzling peculiarities of ants, shared only to so high a degree with termites, is the presence in their nests of quantities of parasitic guests who batten on them, or even sometimes devour the brood or the workers themselves. In spite of this the ants tolerate them completely and even, as we shall see further on, witness the murder of one of their fellows with indifference; the actual victim will allow herself to be eaten without resisting or seeking to escape. It does not usually go as far as that, and other guests are often merely tolerated. Among these are numerous acarines, collembola and little cockroaches such as those that haunt the fungus gardens of *Atta*. Then there are the cheeky silver-fish called *Atelura*, who have the impudence to slip between two ants as they are exchanging food and snap up the sweet droplet as it passes from mouth to mouth. There is even a mosquito that plays much the same trick: it lies in wait beside the track along which throngs of workers are returning, swollen with honeydew, and by tapping some of them sharply on the head it makes them disgorge for its benefit. Still more numerous guests confine themselves to licking the bodies of the unresisting ants.

Bernard found the nests of the desert ants crowded with commensals who in this way had discovered the only damp spot for hundreds of metres around. Finally, we should note that there is a whole series of blind guests, fragile and not very fecund, who seem able to live only in ants' nests; an example is the little cricket *Myrmecophila*, which lays but one egg a year. . . . Other parasites cling to the workers and suck their blood. One of these is the very small beetle *Thorictus forelli*, which attaches itself to the base of the antennae, pierces the integument and gorges itself with blood. Then again there are the *Mermis* worms which live inside their host and appear to stimulate its appetite inordinately; what is more, the ant assumes a form halfway between soldier and ordinary worker.

The number of these guests is incredible. As early as 1894 Wasmann had counted 1,246 species, including 1,177 insects, 160 arachnids and 9 crustaceans. Now more than 3,000 species have been identified. But it was the strange "myrmecophiles" in particular that fired the enthusiasm of the celebrated German Jesuit Wasmann, one of the fathers of myrmecology, distinguished both for the range of his knowledge and his appalling temper (if one may judge by the tone of his controversies with another equally renowned myrmecologist, Wheeler). Bees, hornets and even the social spiders (who are beginning to be studied) also shelter some curious guests (the spiders, for instance, with poetic justice, harbour a fly that devours them!) but there are far fewer than in an ant nest or termitary.

A strange perversion

Nothing can give us a better idea of this curious yet frequent sort of occurrence than Le Masne's description of how *Paussus*, a little beetle with enormous club-shaped antennae, sets about doing the ants the greatest harm possible. It creeps into the nest, where it is accepted without demur, and the workers seem to take a keen pleasure in licking it. *Paussus* appears quite at home, though still hesitant in manner, as if it did not quite know what it was looking for. It feels different objects with its palps, including various parts of the ants' bodies, such as thorax and head, and then lets go almost at once. If it manages to seize the gaster, however, it clings tightly, and an astonishing scene is enacted, before which Le Masne (the calmest and most scrupulous of myrmecologists, and also one of the best) could not conceal his amazement. With its sharp mandibles *Paussus* tears open the ant's belly and slowly begins to devour it. The victim not only forbears to resist; she partly folds her legs and adopts a "nymphal" attitude, as Le Masne calls it. Sometimes she gently pinches the beetle's antennae. Can it be that she feels some fearful delight at the touch of her executioner's secretions? Other ants may pass by while this is going on without making the least effort to rescue their companion; far from it, for some will even lick the beetle. Le Masne never managed to witness the actual death of the victim, who can survive for several days.

Is this monstrous insensibility real or only apparent? We shall probably never know, incapable as we are of detecting any emotion on the ant's impassive leather mask. However, such behaviour is not without parallel. An ant who has lost her entire abdomen can behave, for a limited time at least, as if nothing had happened, even carrying her load of wood to the nest. Other observers have seen a bee sucking up honey while a *Philanthus* (a sort of predacious wasp) was gnawing her abdomen. Then the victims of a praying mantis make no attempt to escape from their slaughterer, even when it has just slain one of their companions before their very eyes. I have myself watched a mouse walking calmly along the back of a viper who was about to kill it a moment later. Even with mammals the apprehension of danger is an odd, irregular phenomenon that manifests itself only in certain circumstances. For example, antelopes are often to be seen drinking or grazing in the neighbourhood of lions who are not in the mood for hunting; then suddenly some imperceptible sign betrays the wild beast's change of mood, and the deer take to their heels with one accord. If we cannot understand an antelope, how can we hope to understand an ant?

With all this, some of the mysteries of *Paussus'* biology remain unrevealed. For instance, they fly with ease, and sometimes several of them may meet together inside an ant's nest. There may be as many as a hundred; do they mate there? We do not know, but surely such numbers could not live permanently in the same nest without murdering all its inhabitants?

The way in which an enemy manages to get accepted is not always as simple as with *Paussus*, as Le Masne and Torossian have observed in the case of another, very different, Coleopteron, *Amorphocephalus* (whose head is perfectly well formed in spite of its name!). The nest it chooses to invade is that of the big ant *Camponotus*, whose temper is not notably long-suffering; in fact it receives a poor welcome and is rudely jostled. Its stratagem is to refrain from the least movement of self-defence or

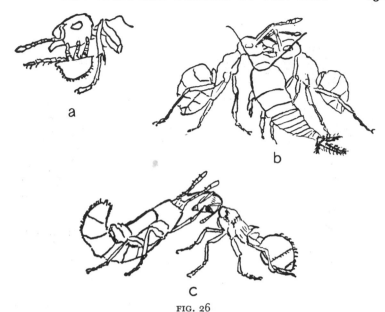

FIG. 26

How nest guests get food: (a) *Antennophorus* clings beneath the head of *Lasius umbratus* and stimulates the ant's mouth parts; (b) *Atelura* steals food during a trophallactic exchange between ants; (c) *Atemeles* taps the head of a worker to induce regurgitation [(a) and (b) after Janet. (c) after Wheeler].

flight, in short, to remain perfectly still. After a few hours the pushing and pulling become rarer and less violent. In any case the violence is never severe enough to cause *Amorphocephalus* loss of life or limb, which is remarkable considering its assailants' air of determination. *Camponotus*' normal ferocity can easily be tested, says Le Masne, by introducing a strange ant to the nest: the attack is instant, violent, remorseless, ending only in death and dismemberment.

Why is the beetle allowed to escape with its life? Torossian thinks that it is because its heavy armour-plating cannot be penetrated by the ants' jaws. Yet the armour would not prevent them from biting off legs or antennae, which they never do; and besides we often see them drag into their nests insects, such as weevils, that are just as strongly protected. In these cases the prey is usually sprayed with formic acid, a powerful insecticide; perhaps *Amorphocephalus* is resistant to it. In any case, the chief

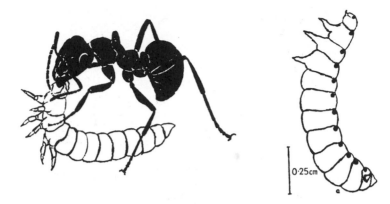

FIG. 27
Formica ant feeding an *Atemeles* larva. On the right, positions of the
attracting glands in *Lomechusa* and *Atemeles* larvae (the glands shown
as black dots). (After Hölldobler, 1967).

reason for its survival probably lies in its immobility, since ants
seldom attack completely motionless prey. Be that as it may,
after a while the creature begins to move and is no longer
molested. By this time, it may have acquired the odour of the
ant nest.

In any case, not only do the ants begin to tolerate its presence,
but they grant its requests for food. Yet more surprisingly,
Torossian discovered, *it gives food to the ants*, a two-way arrangement that seems to be rare among myrmecophiles. Things only
go as far as that with its normal hosts, *Camponotus*, the largest
ants in our region. Everywhere else, even among species of ants
where it is not found in nature, its first introduction provokes
only a short-lived jostling, but it is never fed, even after its final
acceptance. In case of need it can find its own food by licking
honey, eating some of the ants' prey, and so on.

Another beetle, *Atemeles*, is adopted, and almost without
delay, by the red ants, because of its larva's special glands, the
smell or taste of which the workers find delicious. The larva, left
to itself, raises its head and moves its mouth parts very much
like a hungry ant soliciting food from one of its companions: at
least, that is how Hölldobler describes it, although in my
humble opinion the likeness is not very marked! At all events, it
works perfectly, for the ants will feed *Atemeles* larvae sooner

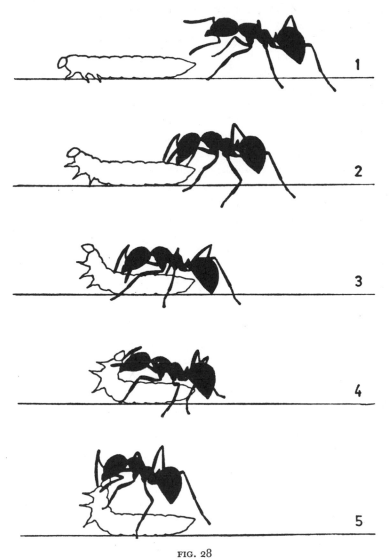

FIG. 28
Formica ant approaching an *Atemeles* larva and then feeding it.

than their own. The ant tackles the *Atemeles* larva in a curious manner, "sniffling" or licking it from back to front (see Figs. 27 and 28). When she reaches the front part the larva rears its head backwards and attaches its mouth parts to those of the ant, so that the exchange of food can begin.

The ants' crazy predilection for *Atemeles* does the community no good, but the damage caused by this unnatural addiction is nothing compared to that of another perversion: the admission of *Lomechusa* to their nests. The name *Lomechusa* is that of an Ancient Roman poisoner, and has always had a sinister sound to my ears. . . . The actual creature is an insignificant-looking little beetle about the same size as an ant. Yet, when it enters a nest the inhabitants are doomed. At first it is greeted with a degree of mistrust, but as soon as an ant approaches with hostile intent, *Lomechusa* presents the hairs on its hindquarters, the trichomes, which carry a highly attractive, and probably sweet, secretion. If that is not enough, it sprays into the ant's face the repellent, even toxic, contents of a special gland. Whatever happens, as soon as the ant has sniffed the two trichomes, all is lost. The workers are no longer interested in anything but the deadly secretion, and neglect their larvae so badly that they produce deformed adults. During this time the *Lomechusa* lays its own eggs in the midst of the brood, and the ants will look after them to the best of their ability. As soon as the *Lomechusa* larvae hatch out they begin to devour all the ant eggs and larvae within their reach. After a short period, addiction to the trichomes' perilous juice produces various symptoms of intoxication in the ants: they lose their sense of balance, for instance. From that time on complete degeneration comes quickly. It does not matter to the *Lomechusa*. When the enfeebled colony is close to death, they leave the nest to seek other victims.

I would like to stress this "lomechusa-mania" among ants as *perhaps the only known case of drug addiction among animals.*

Ant commensals and parasites of ants

Sometimes one species of ant will form with other ant species relationships of a most curious, paradoxical and almost unbelievable nature.

Myrmecologists have long been aware that more than one species of ant may be found in the same nest. The simplest case is that of so-called compound nests, as opposed to single nests. Here we should pause for a moment to mention the rather extravagant names with which the more richly imaginative myrmecologists have endowed the various types of associations (plesiobiosis, leptobiosis, xenobiosis and so on). There are, for example, triple associations (*Colobopsis truncata, Dolichoderus quadripunctata, Leptothorax tuberum unifasciatus*), observed by Stäzer in the hollow stalks of bramble bushes while he was studying this subject. Very often the nests of these species intercommunicate; at the very least their entrances and exits are common, arguing the absence of aggressive reactions between them. Such cases may well be called *cohabitation*.

There also exist cases where the cohabitation shades off into hostility. One such situation concerns the minute *Solenopsis fugax*, who establishes its colony in the thickness of the gallery walls of much larger species. The relationship is definitely hostile, the more so in that *Solenopsis fugax* is very aggressive and does its best to plunder its hosts, while the latter are too big to pursue it through its tiny corridors. A third sort of association may be closer, yet without hostility. An example was that found by Wheeler between *Megalomyrmex* and *Sericomyrmex*, the fungus-growing ant. They live together and even lick each other. They both share the benefits of the fungus gardens, but only *Sericomyrmex* cultivates them.

Slavery

The development of ant associations follows a sort of cruel logic, and the last-mentioned case is already halfway towards the next stage, slavery. The tendency to subject alien species to one's service, so painfully reminiscent of *Homo sapiens*, is a habit that has made ants famous, but hardly more appealing. Stumper has pointed out one odd fact, difficult to explain: slavery is only met with in the *temperate zone, never in the tropics*. In Europe two families are especially known for this type of behaviour, *Formica sanguinea* and *Polyergus rufescens*. The *sanguinea* female, once fertilized, is unable to found a new colony alone. Therefore she enters a *Formica fusca* nest to collect around her as many pupae as possible. She kills any worker who tries to oppose her. Indeed, surprisingly to anyone who knows the aggressiveness of ants, they do not attack her for long. Soon the *sanguinea* queen, alone behind her barrier of pupae, presides over the hatching of the first *fusca* workers. They begin at once to feed the *sanguinea* queen, and tend her eggs as if she were their own queen. Later the young *sanguinea* ants, while continuing to take over the *fusca* pupae in the nest, *go out to abduct more from other nests.*

People wondered how this "kidnapping" behaviour could have arisen among *sanguinea*, whose old queen is as timid as any other ant queen—yet it is she alone who could have transmitted that hereditary characteristic. The puzzle arose because only old *sanguinea* queens had been observed. Place a *young sanguinea* queen among pupae and workers of a slave race, and she will, as we have seen, display extreme aggression, and seek unremittingly to filch cocoons and kill the workers. But as soon as the first kidnapped pupae, which she has piled up in a corner and guarded jealously, begin to hatch, her behaviour changes completely: she becomes very timid and avoids the light. Thus we see that the adult queen's aggressive nature is only masked, but shows plainly enough at the beginning of her career, as it does all the time with the workers.

Miss Fielde (one of the very few female myrmecologists) has shown that the case of *sanguinea*, who steals the pupae of other species, is possibly not as aberrant as one might first think. She managed to make three or four different species, such as *Stigmatomma pallipes, Formica subsericea* and *Aphenogaster fulva*, all live together. They all licked each other and exchanged food without distinction. These unions can be formed up to twelve hours after emergence from the cocoon simply by putting the young insects together; but if this is delayed beyond the critical moment, hostility develops rapidly.

With *sanguinea*, however, slavery is not indispensable. It is true that they cannot found a colony without *fusca*'s help, but this stage once over they could exist perfectly well without making any more slave raids. The case is very different with *Polyergus*, an ant with long and pointed mandibles, suitable for war alone and not for work. She cannot even feed herself and is forced to make periodic raids on *fusca* pupae for her own colony to survive. These ants make war very methodically, probably sending scouts on ahead. Subsequently all the *Polyergus* leave the nest, hurl themselves against the nest selected, and emerge soon afterwards each carrying a pupa in her jaws. The fertile *Polyergus* females also get themselves tended by colonies of *fusca*.

The "scouts" and the "captains"

But Dobrzanski and Dobrzanska maintain that we must not believe Forel's assertions that the *Polyergus* expeditionary forces follow the scout's directions, and are led by "captains". That idea is too anthropomorphic. The truth is stranger still, and fits better with what we already suspect about ants, that though they may behave like humans, it is only in outward appearance; their methods, in reality, are utterly different. Some hours before the raid, isolated individuals leave the nest and run a certain distance forward. These are the supposed scouts, but one soon observes that the main wave of Amazons does not flow in the direction they have taken; nor does removing all the "scouts" hinder the raid. One hour before the raid, however, other, particularly agitated, individuals emerge, called by the Dobrzanski the "activators". From time to time they may return to the nest, where their presence creates a rising excitement, until finally the main body of workers rushes out in the direction of the mass of "activators": *to remove the latter is to suppress the raid.* But they do not march at the head of the column, although well towards the front; the leaders change constantly. The column seems to advance straight though at random. If pupae are thrown in the path of the workers, they do not pick them up. They must find them for themselves in the nest.

The slaves help their masters to enslave others

How do the workers hatched from the stolen pupae behave afterwards? It has long been known that they serve the needs of their captors. But some very remarkable observations made by Wilson (1965) show that the collaboration goes further than that. He was studying the *Formica* slave-makers, who are not particular as to what kind of ant they enslave and will willingly take slaves from several different species. Now these slaves retain at least some of their own characteristics. For example, when *Formica sanguinea* has abducted *Formica pratensis* pupae, the workers that emerge tend to make nests somewhat like those of their own species. Forel actually got some *sanguinea* ants to adopt pupae of the Amazon *Polyergus*; once hatched the young *Polyergus* firmly refused to do any work at all, quite in the tradition of their family.

Under the heading of *modifications of slave behaviour* through contact with the master, I must cite the strange case of *Lasius umbratus*, the exclusively subterranean ant, which is interested only in root aphids. When *Dendrolasius* calls on it to found a new colony *umbratus* will accompany the first *fuliginosus* workers to the treetops in the full light of day (Verhagen, 1930)!

Multiple slavery

In the United States Wilson had the good fortune to find an ant (*F. wheeleri*) which had captured at one and the same time pupae of two separate species, *F. neorufibarbis* and *F. fusca*, both with very different behaviour patterns. Not content with that, the *wheeleri*, at the time Wilson had them under observation, were busy stealing the pupae of yet another species, *F. lasioides*, all in the course, moreover, of robbing a nearby *fusca* nest. They were accompanied by numerous *neorufibarbis* slaves and a few *fusca*. It was the *wheeleri* who seized and transported the captives, the *neorufibarbis* helping only by tunnelling new entrances into the besieged nest. Normally it is unknown for *neorufibarbis* to make raids, and it is very extraordinary that they should have arrived at such a collaboration with their masters. It is in direct contradiction of what I said just now about slaves keeping to their natural behaviour in their masters' nests, and implies a great adaptability of behaviour in *neorufibarbis* and possibilities of evolution one would never otherwise have imagined, apparently quite lacking in *fusca*. The latter are concerned only with feeding the larvae of the master race, and storing food in their own gasters. One finds plenty of them waiting about the nest, quite still, with swollen abdomens, in the classic "replete" attitude (from the Latin *repleta*, full), ready to exchange what they have stored with the rest of the colony. Could a colony of slave-makers recruit in this way several specialists of different vocations? That is going a long way indeed, and climbing high, nearly as high as mankind, on the evolutionary ladder; a ladder parallel with mankind's, so to speak, where apparent likenesses hide a profound difference of mechanisms. We will come back to that. I should also like to know whether all *wheeleri* are capable of this multiple recruitment, or whether it is the privilege of a few innovators. The question is not as absurd as it might have sounded some years ago. Innovators have been found, for instance, among apes and birds, but as yet no one has had the idea (or the courage?) of looking for them among ants.

How did slavery develop?

Here we are obviously reduced to rather vague guesses. When Gösswald was studying this question he began by trying to lay down a classification, distinguishing, for example, between those predacious colony founders that kill the host queen and those that do not. The murder is likely to happen when the workers of the host species are able to reproduce themselves, or when the predacious ants can procure a reserve of workers, as when *Polyergus* go on pupa-stealing expeditions. If these two possibilities are excluded the predacious queen has no choice but to leave the host queen alive. This is certainly the case with *Strongylognathus*, which forms a veritable alliance with *Tetramorium*; the two queens live together for a long time.

According to Gösswald, the founding of a nest by means of slavery is advantageous to ants since it avoids the enormous loss of queens: for instance non-slave-making ants will be lucky if one queen out of five hundred succeeds in establishing a new colony. But to tell the truth I do not see how the young "guest" predacious queens should avoid the hundreds of mandibles, fangs and beaks that lie in wait for them during their nuptial flight any better than the queens of their victim species. Gösswald puts forward another slightly more convincing theory. At the time of mating a large proportion of the eggs are no sooner laid than devoured by the queen, who in many cases lives in a closed circuit and cannot get food for herself. In the contrary case, when she is helped by host workers, many more survive to the hatching stage. Nevertheless, we must admit that we lack sufficient accurate information as to the advantages and disadvantages of this parasitic guest life.

As to any particular anatomical peculiarities that predispose the females (queens) to adopt this parasitic life, nothing certain has been put forward. Wasmann noted the small size of all parasite queens. But there are a number of independent colonies where the queen is no larger, in *Myrmica*, for example, where the queen is scarcely larger than a worker. Others have pointed to the influence of climate; Gœtsch and Käthner found that *Camponotus chilensis* lived as a parasite in the colder southern regions of Chile, whilst in the hotter northern area it led an independent life. Wasmann found that under our climate conditions all the parasitic ants were northern. But we cannot say anything definite yet as regards parasitism and temperature.

True parasites

We now come to the true parasites. One of the first to be studied by the great Forel was *Strongylognathus*, which is no longer able to abduct and lives solely at the expense of a very common species of ant, *Tetramorium caespitum*. Other ants do much the same with other species, except that the invading queen begins by killing her rival (see Fig. 30) or, refinement of atrocity, seems, by heaven knows what mechanism, to *induce the workers to kill their own mother*. True parasites are not even capable of an independent life: as Stumper says, there appear "marked characteristics of adaptation and degeneration in both males

FIG. 29

A male and two females of a *Teleutomyrmex* species parasitizing a female of a *Tetramorium* species. One of the female parasites is physogastric, that is its abdomen is swollen with food reserves and eggs. On the right, beneath, a *Tetramorium* worker (after a drawing by Linsenmaier).

and females; atrophy of the mouth parts in both; a 'larval' form, either apterous or not, in the male; adipose hypertrophy of the fertilized female's abdomen (physogastry) with an abundant secretion from the cutaneous secretion glands that the host workers find very delicious". Add to all this the disappearance of the worker caste: only the sexuals remain, and they, like the *Teleutomyrmex* (see Fig. 29), live permanently attached to the body of the host queen.

Food exchange between host and parasite

It is also possible to measure, with the help of isotopes, the exchanges between host and parasite. This has been done by Gösswald. We must, for instance, distinguish on that head between *sanguinea* and *Polyergus*, the best known "Amazons". Their slaves give more food to the first than to the second, doubtless because *sanguinea* looks after her own brood, feeding it with her regurgitations. Thus *sanguinea*'s need of nourishment is greater than that of *Polyergus*, who leaves even the feeding of her young to slaves. *Polyergus* never of herself takes the radioactive food placed in the container, nor is she ever seen practising trophallaxis. *Sanguinea* ants, on the other hand, can perfectly well feed themselves, and independent colonies of this species are found which seem no less prosperous than those provided with slaves. Very degenerate parasites, such as *Epimyrma* and *Anergates*, never feed themselves at all. The isotopes also show that the beetle *Lomechusa*, whether in the larval or the adult state, receives plenty of nourishment from its hosts; from

FIG. 30
Female *Epimyrma* (black) about to cut the throat of its host's queen, *Leptothorax nigriceps* (after Linsenmaier).

sanguinea, that is to say, but not from its slave *Serviformica rufibarbis.* This is because *Lomechusa* is a *specific parasite of sanguinea.* We also learn from isotopes that the former conveys divers substances to *sanguinea* ants, who persistently lick it and thus absorb various secretions. In the same way, if a radioactive *Atemeles* larva is placed amongst hungry ants, they will become radioactive themselves by licking it, and it can be shown that the active substances filter out through the dorsal glands of *Atemeles.* One can even wash the larvae in acetone and impregnate a piece of filter paper with the extract; the ants will carry it to the brood-chamber, where it will be fondled as if it were indeed an actual larva. In this case the chemical stimuli are so powerful that they overcome the tactile stimuli; for the organs of touch must make the ant well aware that it is not dealing with a larva . . . yet it makes no difference.

CHAPTER VI

SOCIAL RELATIONS

Methods of communication — Alarm — Exchange of food

Do ants "talk" to each other?

MANY AUTHORS HAVE wondered by what process ants inform their companions that food has been found: for there are among ants means of rallying workers quite as effective as those that exist among bees: but ants have no dances, though possibly a scent message, as found among the *Meliponae*. In the case of the red ants of our own woods, the forager who finds food first gorges on it, then comes back to the nest and offers it to her companions, with no preliminary gestures or rubbing of antennae: she merely presents the droplet hanging on her mouth parts. If the colony is hungry the workers will gather round her forming a rosette of ants and take their share of the food. Then either they in their turn distribute some of it to others or themselves leave to forage. They do not seem to give any indication of direction or distance, as do bees, but they manage to spread a general, diffused excitement.

In the case of *Solenopsis* an important additional factor is the leaving of an odour track, as we shall see: but as far as is known there is nothing like this in our red ant. On leaving the nest they find only the network of tracks with the rush of traffic on it something like a railway marshalling yard.

That is the classic theory: I must admit I find it difficult to believe that things are as simple as that. In fact all the observations were made under laboratory conditions, which are all very artificial. There are a few field observations which give some additional information, throwing a rather different light on the subject. We find (1) tracks, 100 metres long at times, which are remembered by the ants from one year to the next, as Wellenstein has shown, and (2) some basic specialization, for instance, the ants have foraging teams, each forager being concerned with a particular tree to which she remains faithful for months at a time.

Scouts, who find new hunting grounds or new aphids to tap, can be found among ants as well as among bees. In the case of *Monomorium pharaonis* every discovery of food is preceded by the departure of only five foragers per minute, and even less with *Myrmica rubra*. They seem to be old workers, because they are darker and show some signs of physical decay. When a scout has found food she brings it back herself and returns to the nest (leaving a scent trail) sometimes keeping a straight line for more than two metres; then, throwing herself into the nest, she rubs against the other workers, tapping them with her antennae and legs. After that large numbers leave the nest, sometimes accompanied by the queen, to find the prey. The number of recruits a single scout can raise varies greatly according to species: very few with *Monomorium*, nearly fifty with *Myrmica*. Much better results are obtained if the colony has already been stimulated by a previous discovery, even if it happened fifty minutes earlier.

It has been suggested that the number of foragers a scout collects is proportional to the size of victim discovered, but there is no good proof of it. This is the case with bees and it might also occur in ants: there would be nothing untoward in supposing that the discovery of a large food source excites the scout more and that when she returns to the nest she rubs and taps the companions who can help her more vigorously. And there is no reason for supposing that their language is entirely different from that of the bees. What we lack are careful observations on a sufficient number of ant species. In any case, with *Pheidole crassinoda* the recruits are three or four times more numerous than is needed to bring back the prey: but those whose services are not required lose no time in getting back to the nest.

Is there an "antennal language"?

I have never quite believed in the casualness of the tapping of antennae affirmed by some authors, particularly since Montagner's recent authoritative work on wasps, which are close to ants in certain morphological and physiological respects. With wasps, feeling of the antennae accompanied by patting with the forelegs, apparently meaningless, can be observed. Montagner took films of them, not without some difficulty: and after hundreds and hundreds of observations he was convinced that the tappings were not at random but were "coded", that is to say the "dominant" wasp (the one that will lay before the others, eat the later layers' eggs, and ceaselessly demand food from her companions) makes her "social status" known by a particular posture and a certain method of antennal tapping.

This is a very important discovery, all the more so as the antennal code is possibly not limited to that alone. But Montagner was favoured from the cinematographical point of view by the wasps' colouring—yellow, red and black on the white background of the nest. Moreover, the antennal stroking is comparatively slow. But with ants it's another matter. How on earth can we record anything amid the wild turbulence of thousands of over-excited workers against the brownish background of the ants' nest? Let us hope that an improved technique will let us do it one day.

In any case, the problem of communication of ants among themselves can, it seems, only receive individual solutions, different for each species. For some little time now we have known that a certain amount of information must be imparted by means of the touching of antennae, but there are only a few precise descriptions in the literature. I will quote some of them.

The *mechanics of food exchange* in *Formica fusca* have been studied by Mrs. Wallis. She has not been able to identify the exact stimuli, or system of stimuli, that release the food-exchanging impulse. However, in trophallaxis there are some gestures which are easily recognized and always the same. Sight does not play a part, which is very understandable as most ant activity takes place in pitch darkness inside the nest. It is more a matter of a combination of complex movements with the antennae and forelegs.

To begin with the ant seeks to find out if it is in contact with another ant, which it does by touching the latter's body. Then the head must be identified, which probably occurs the moment

FIG. 31

Above: trophallaxis, a fundamental basis of social life, the exchange of food between two workers of *Formica fusca*. *Below:* cleaning activities of *F. fusca*; (a) Cleaning the right antenna and the left anterior leg; (b) Cleaning the other legs; (c) Cleaning the gaster (after Wallis).

contact is made with the antennae. Touch is not the only sense used; most probably smell comes into play too: antennae with their chemotactic sensors must register it and ensure the correct position vis-à-vis the other ant. Movements of the palps seem to speed up regurgitation, but they are about the same in both donor and receiver. Nevertheless, the acceptor seems to stroke the head of the donor to a greater extent, often using the fore-legs. The giver makes much less effort to direct its movements and is content to stroke the head of the acceptor somewhat absent-mindedly.

As to what makes an ant a donor or acceptor, it seems to be merely a matter of hunger. If one may believe Wallis, the hungrier worker solicits and accepts food. It is not so much a matter of individual but of collective hunger: in a hungry colony the number of ants soliciting food increases. As to the donor, there is no doubt that the extended crop, being full of food, leads the insect to offer it to others.

Wallis also tried comparing the behaviour of two ants, *Formica fusca* and *F. sanguinea*, of which the first is the slave of the second, when introduced into a strange colony. *Fusca* tries to escape and runs about with irregular movements all over the place. On the other hand *sanguinea* taps the antennae of the ants it encounters and even tries to offer them food. This reminds one of the behaviour of the beehive sentinels: these old workers take up positions near the entrance and carefully examine every entrant. A stranger straying in is quickly examined, immobilized or even stung, and in any case turned out. Unless, that is, the intruder has time to offer food to the sentries, adopting the accepted pose for this purpose: the sentries, it seems, are easily corrupted!

But it cannot be said that it is also an attempt at conciliation in the case of *sanguinea*. The antennal strokings do not at all disarm the assailants of the stranger colony, and most of the time the offer of food is not accepted.

Other phenomena involving the use of antennae

Szlep and Jacoby have recently studied one of those pheno-mena known for a long time but never explained. It concerns tremblings running along the body, while the head is pushed forward in jerks, the antennae shivering and the whole body vibrating. In *Myrmica*, *Tapinoma* and *Tetramorium* these are undoubtedly "recruiting movements": a sort of dance, somewhat comparable with the bees' round dance (remember that this dance is not directional and is thus different from the figure-of-eight dance; the round dance stimulates the bees to go out and look for food). This trembling occurs in ants after the discovery of food by a scout: its effect is to cause a number of foragers to leave the nest, among whom the scout may not necessarily be found. They spread out at random over the trails: and it may well happen that some, or even all, the ants turn back to the nest well before finding the food. This is a strange aspect of the ant's social life, which we shall discuss in the final chapter. Often, also, the ant discovering food distributes it around her.

We must also mention a primitive but efficient type of alarm mechanism that is found in *Cardiocondyla*. It is called "tandem communication" (also found in some *Camponotus*). An ant, having found food, returns to the nest empty and in a very short while comes out with another ant behind her feeling with her antenna the first ant's abdomen; this stimulates the discoverer to advance a few centimetres and then to stop, to allow the second ant to catch up with her, who then again taps the end of her abdomen with her antennae. It seems that this is the start of the process of trail-marking found fully developed in *Dendrolasius* and *Solenopsis*.

Preferences in food exchange

Gösswald and Kloft have used radio-isotopes to study food exchange. An ant fed on labelled sugar shares it with from eight to ten others at most. These will redistribute it in their turn, so that at 25°C, the normal nest temperature, the crop content of one radioactive worker can be spread over eighty others. But the exchange is very irregular: for instance a worker who has just given out food may herself solicit it a few moments later from the very ant she has just fed! First the *majores*, then the *minores* and finally the *minimae* take part in these exchanges. When young workers want food they solicit it more from old workers than from young ones. If the young are isolated as a group their ovaries stop growing even if they are getting abundant food. Thus normal development needs an exchange of food with older workers, who must possess some substance the young workers do not yet have. It is a strange phenomena, not yet explained, and one that does not appear to occur in bees, where a group can be made up consisting only of young bees, and yet their development seems to be entirely normal.

Stimuli such as the smell of the queen or the nest play a great part in trophallaxis. If two groups of ants are starved for forty-eight hours and one group (A) allowed to acquire the nest smell through a wire screen, whilst the other group (B) is completely isolated, then, on returning the two groups to the nest, group A (with the nest smell) will be the first to be fed. But if Group B has been kept in the presence of a dead queen, B will be the one fed first, before A, the nest smell group. This is normal practice in food distribution. The foragers do not discharge the contents of their crops directly into that of the queen but into the crops of the internal service ants, who then in turn give it to those closer and closer to the queen; finally the queen gets nothing but the contents of her attendants' feeding glands.

The big idlers

A large number of big, idle workers is found in the nest; they do absolutely nothing all their life except take part in food exchanges, and the foragers, it is said, give preference to them over all the others. In any case one never sees these idlers do any work at all: perhaps they are an essential link in the social feeding chain or are a food store.

Schneider does not think this involves any privilege for the big workers. According to him, it is only the intensity of demand that regulates exchanges of food and their duration. The hungriest workers solicit food more persistently. It can happen, as we have seen, that in a more or less full-fed colony "double exchanges" take place; two ants solicit food from each other at the same time and both disgorge a droplet.

Food segregation

Some authors, such as Gœtsch, have noted separate "food groups" in *Lasius flavus*, and nothing seems to be exchanged between these groups. For instance, if you mix trypan blue dye with the food of one group and a slow-acting insecticide with that of another it will be found that none of the dead ones have any blue in their crops and *vice versa*. Food distribution from the crop, then, is not made to all the inhabitants of the nest, but only to ants within a group. Moreover, Gœtsch has found these closed groups in other species; on the other hand Otto has proved that *Formica* workers exchange food with each other indiscriminately.

Corruption of the staff

Another essential role of food exchange is the overcoming of hostility, greater or lesser as may be; for example, when some strange ants are introduced into a nest great numbers of the intruders open their mouths and regurgitate a drop of food, even if they are hungrier than the legitimate occupants. If the proffered droplet is accepted, it is a sign that they are going to be too. We have seen that this is very common in bees (see p. 147).

Zahn has added another important observation, that trophallaxis has the effect of equalizing water content, a substance easily lost by ants. An increase in food exchange is found as soon as the temperature rises.

Two-way exchanges

Finally, we must note that exchanges are mutual between different stages and castes: not only from nurse-ants to larvae but also *from larvae to the nurses*. Gösswald proved this by offering radioactive larvae to normal nurses: the nurses later became slightly, but nevertheless measurably, radioactive. These radioactive isotopes have also proved that the queens secrete from the thoracic glands a substance greedily licked up by the workers. Maschwitz has shown that larvae play the part of protein reserves: unfed workers live longer when with their larvae.

Chemical language

Wilson found no less than seven olfactory methods of communication in *Solenopsis saevissima*. They are:—*Nest smell*, this does not come into play unless some extraneous factor is introduced; *body odour*, which can be extracted by a solvent and can then give rise either to "clustering", a form of swarming, or to cleaning operations; *Dufour's gland secretion*, which is an attractant and serves to odour-mark trails, or to assemble companions during a fight; and, finally, the *secretion of a substance at head level* which is an *alarm* signal and gives rise to considerable agitation among the workers. In addition there are other substances with less well-defined roles.

A special mechanism is used for *marking trails*. The Dufour gland first produces a colourless secretion which flows along the sting, the ant touches the track at regular intervals along the way with the end of this organ, as if she were drawing a "dotted line". The marks cannot be seen, but they greatly influence ant behaviour and lead the ants to follow the marked path. That is all; no other information is given, such as the quantity of food available or even its exact location. In fact the ant that has found food automatically brings out its sting and marks its return path: this return track can be most irregular and even make several unnecessary loops. However, this apparently crude procedure has some curious results, as Wilson found out, and which I will describe in order to show that though behaviour in ants is often co-ordinated and apparently like that found in man, it is in reality effected by non-human means.

It was noted above that the larger a food source is, the more ants gather round it. At first sight one might think ants pass on information on the size of the food supply, as do bees. In fact, after the ants put down their dotted line of odour, it evaporates in the course of a few minutes. If the food is not of much size, and if but few ants go to it, then only a short time will elapse before the substance evaporates and the trail is no longer marked. On the other hand, if the food is swarming with ants, new arrivals cannot get to it and they return to the nest without marking the track; thus the number of ants at the source depends on its size, without any exchange of information from ant to ant being necessary. The marking odour is specific to any given species of *Solenopsis*.

Obviously, if these markers evaporate so quickly, they would not appear, at first sight, to be capable of guiding workers over long distances, say of more than two or three metres. But, in fact, ants easily track down food sources at distances of this order. Wilson tried to explain this in what I think was an involved and unconvincing manner, by the tendency of *Solenopsis* to overrun its mark and to continue travelling in the same direction.

However, *Solenopsis* explores the neighbourhood, as do all

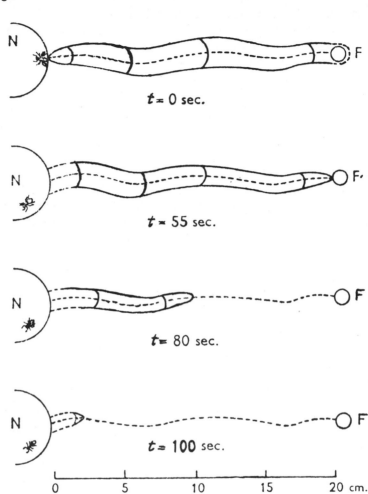

$t = 0$ sec.

$t = 55$ sec.

$t = 80$ sec.

$t = 100$ sec.

FIG. 32
The odour track left by *Solenopsis* from the food (F) to the nest (N). It
is a band 40 cm. long by 2 cm. wide. Beyond this area the smell is too
weak to attract workers (after Wilson and Bossert, 1963).

animals, and it has been observed that it marks the unexplored areas in the neighbourhood of the nest as well, even if it finds no food source, which is hard to understand. According to Wilson's calculations, though the method of passing this information is very different from that of the bees' dance, it is no less effective if measured in terms of information units ("bits").

Finally, Wilson showed that it was Dufour's gland and not the poison gland that was responsible for the track-marking. He used crushed glands of both kinds to mark tracks and only Dufour's gave paths easily followed by the fire ant.

The alarm substance

All beekeepers know that it is not wise to crush a bee when handling the hive: you greatly increase the risk of being stung. Up to 1964 everyone believed that the alarm substance was the poison itself. But Maschwitz had the patience to dissect out each of the glands found round the sting; the Dufour gland, Kochevnikof gland and the sting sheath glands (newly discovered). He found that the poison itself had no effect. The alarm substance is a special oily secretion exuded between the plates near the sting; however, these plates have no glandular structure behind them; and on the other hand the known glands, if crushed, have no effect when offered to bees. Briefly, we now know where the alarm substance is excreted but not from whence it comes.

In ants the problem is more complicated on account of the many species and kinds of behaviour. Moreover, what exactly do we mean by alarm in ants? Rarely, as in other insects, a flight reaction; mostly a swarming towards the zone where certain substances have been produced, followed immediately by attack. Maschwitz made a profound observation on this subject: given the presence of brood and a queen which must be preserved at all costs, flight makes no biological sense, no more for ants than other social insects. *Attack is the only recourse.* And this is probably the reason why, alone among insects and probably all animals, social insects do not run away from man. Only if the alarm is raised outside, when, for example, ants are eating their prey, does a flight towards the nest take place. And even that only happens with timid kinds, such as *Lasius* and *Tapinoma*. There are some much more pugnacious species, such as the red ants of the woods, *Formica polyctena* and *F. rufa*. These species never fly away but spray formic acid at a high concentration on the trouble-maker and try to bite him. Such tactics can be most effective: I have already mentioned an enormous ant-heap that one of my assistants and I took it into our heads to tease one day (of course with a scientific end in view) on the High Koenisbourg slopes. The nest was taller than us and must have been well over a cubic metre in volume. We had to turn tail for, in a few seconds, the ants covered us with an angry army, half stifling us with their spray of formic acid. The stimulus to fight can be so great that the ants even bite each other: Maschwitz saw this happen when he cunningly smeared their food with a concentrate of the alarm substance. In addition, when ants are trying to overcome a struggling victim, they spray formic acid on it, immediately attracting more workers, who then lend a hand.

But just what are these alarm substances? Their chemical nature and the organs of origin are very variable. I have just mentioned formic acid, the red ant's well-known secretion; the acid was formerly extracted commercially from them. In itself it poses more than one physiological problem: it is stored in a vesicle whose volume is up to one-fifth of the abdomen and is found at the extraordinary concentration of 50 per cent (repeat *per cent*, not per thousand). How is it that the organs making and storing it are not irreparably burnt? The old authors thought the poison was vinegar, and acetic acid does also act as an alarm substance for red ants, though it is much less effective than formic acid. Nevertheless, a solution of pure formic acid is not as active as the mixture of substances secreted by the ant's poison gland; and it has been known for a little while that the secretion from the small tube of the Dufour glands, attached to the poison gland, considerably reinforces the action of the formic acid. According to Maschwitz, there is yet another substance that gives the alarm; its role is not very well-defined and it comes from the mandibular glands, consequently from the other extremity of the body.

The above is the case in all stinging ants, such as the red ants (*Myrmica*). In some families, such as the *Dolichoderinae*, the poison gland is atrophied and a *special anal gland* replaces it for the production of the alarm substance.

In some species the queens also produce alarm substances, though they do not take part in fights, but flee from the scene of battle. As far as is known the male sexuals secrete no such substance.

It must be noted that these compounds *are not strictly specific*: they may work with several species of the same genus and even on different genera. Wilson and Pavan showed that the secretions from the anal glands had the same effect on three different species, though they all smelt very different to man. But in two cases at least chemists have isolated identical compounds.

It should also be noted that no alarm substance is produced by the small colonies of primitive ants (Ponera), which have at most a hundred individuals. This is no doubt due to the fact that any attack or disturbance is felt by the whole colony and there is no need for the transmission of information by means of chemicals.

Territory among ants

Most wild animals, and even some domestic ones, actively control an area of varying size around their nest or lair: it is their *territory*. The idea of territory is inseparable from that of hierarchy, and over the last twenty years these two concepts have revolutionized the science of animal behaviour. Formerly, there was a tendency to think of unorganized hordes of animals moving around at random: this was the "state of nature". Now, nothing is more false. None of them seeks its food where and when it wishes over a limitless area; on the contrary, the frontiers are narrow and well defined and each creature has a good idea of its social position; it is either dominated or is dominating. In addition even a *leader*, or, as it is put, the *alpha* animal, runs the greatest danger of being dominated and severely beaten outside its own, or its herd's, territory.

Ants are no exception, but they add to their territoriality the inevitable complications of their social behaviour. Brian observed several species on the moors in Scotland; he noted that not only the nest but even the aphid shelters (those earthen fortresses that some species of *Lasius* build round their cherished cattle) were part of the territory and were energetically defended. When some ants are offered a spoonful of syrup it might be thought that the rule would be "first come, first served", particularly if the discoverer is capable of quickly summoning a crowd of companions to the prize, though this capacity for rapid recruitment is not found in all species of ants. The behaviour of *fusca* is quite different according to whether the insects are inside or outside their territory. In the first case the ant will not hesitate to attack the stronger *Myrmica* if these dare to approach; but in the second instance *fusca* will flee before the aggressive *Myrmica*. Often there is no fight because of the varying habits of ants; *fusca*, for instance, tends to care for aphids on the tops of plants and *Myrmica* to cultivate those at its base; under these conditions a plant can be exploited peacefully by the two species.

Brian and Talbot have noted that territories used by different colonies of the same species interlock one with the other like the pieces of a jigsaw puzzle, but that they may be considerably encroached upon by territories belonging to other species. As to the size of territories, they vary enormously; from 2·33 square metres in *Myrmica*, 5,000 in *Formica nigricans*, a hectare and more in *F. polyctena* and several hectares in the case of *Myrmecia gulosa*.

However, *it appears that only a minimal portion of these territories is used assiduously,* namely the part adjacent to the tracks used by the ants. They do not seem to be able to scent food at more than a few millimetres' distance and the movements of a possible victim are not noted beyond about ten centimetres. Moreover, as we shall see below, the tracks regularly lose workers along their length as they go off to forage, and this loss is 1·5 per cent of workers per metre. This is balanced by a return to the track of successful or unsuccessful foragers at the same rate. It also appears that hunters are less assiduous in following the tracks than are the honeydew collectors, which is understandable.

I should add, in passing, that the distance to which ants will go to exploit a food source, and thus extend their territory, has nothing to do with their size and species. *Fusca* may go two-and-a-half metres away from the nest, whilst *Myrmica* never ventures more than a metre. By contrast the minute *Leptothorax*, much smaller than the above two species, easily runs out to three metres. As to the red ants, they do not flinch from making journeys of more than a hundred metres, if we may believe Gœtsch. The fungus-growing *Atta* goes even further.

A sort of game

Sometimes the fight between the legitimate owners of a nest and intruders is not very fierce; it depends on the species involved. It can happen that the conquered or "dominated" ant is seized by the petiole and carried into the nest of the victor, where it runs a grave risk of being torn to pieces. On the other hand we have the curious observations made by Brian on some *Myrmica* from laboratory nests: when he put a strange ant of the same species among the ants of one of his colonies the newcomer was firmly grasped by the petiole, giving rise to the immobilization reflex in the victim, who was then carried not into the nest, but to the "cemetery" or rubbish dump outside, where the creature quickly recovered and made off. Brian had no hesitation in interpreting this as "a strange method of reducing deaths by converting war into a sort of game". This is very striking, because we know that the activity of carrying away bodies is released by certain chemical stimuli coming from tissue decomposition. This would not occur in the case mentioned by Brian. We shall not properly understand this strange experiment unless we repeat it. If it really is a kind of game, we must note that up to the present it has been an accepted truth that insects do not play. The old authors, however, have always reported almost play-like behaviour—and in ants, too. . . .

Frontier incidents

In any case, in nature, and within a species of ant, a kind of alliance seems sometimes to exist. Brian found five or six *Myrmica rubra* nests living peacefully together in the same rotten tree stump: the foragers from each nest all travelled in quite different directions.

The other behavioural extreme is found in the tree *Œcophylla*, where each nest occupies a tree on its own and will tolerate no intruders. Nevertheless, another ant, *Anoplolepis*, is not intimidated by the *Œcophylla*; it does not live on the tree itself but inhabits a subterranean nest at its base. When one of the fierce weaver ants passes by, *Anoplolepis* simply seizes it and then cuts it up inside the nest. This is yet more strange in that *Anoplolepis* is not particularly pugnacious, and tolerates many other ants, some of them very big, on its territory. This, no doubt, is due to the fact that these ants are no threat to it, whereas *Œcophylla* attacks at once.

Finally, Wheeler long ago noted the peaceful visits made from nest to nest along the connecting paths. Talbot even found some workers he had marked in one nest helping build another one.

Obviously there would be nothing odd about this in poly-calic colonies since they all come from the mother nest with which the daughter colonies have exchanged brood, queens and food for a long time. However, Scherba has noted ants visiting a *Formica opaciventris* nest which was not polycalic in origin. The reason for such visits is by no means clear.

Ants as chemical factories

We have already noted, on several occasions, that ants secrete the strangest and most caustic of substances, which stop their enemies coming too near and usually give an advantage in fighting. The wood ant's formic acid is a highly effective weapon.

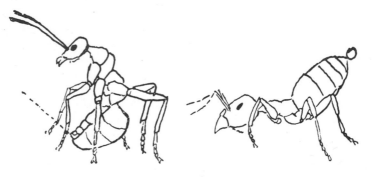

FIG. 33
Left: Formica worker spraying poison. *Right: Lasius niger* worker in state of alarm; a drop of venom is at the extremity of the abdomen (after Maschwitz, 1964).

Recently chemists have taken a great interest in insects precisely because of the presence in their secretions of certain mystifying substances. The chemists are unable clearly to explain how it is that these substances do not destroy the very tissues producing the secretions. I know very well that our own stomachs produce hydrochloric acid without our being affected, and we more or less know why. . . . But all the same, the hydrochloric acid is not almost pure, at 100 per cent concentration; it is relatively dilute. When the caterpillar of *Dicranura vinula* spins a cocoon in which to pupate, the insect makes it almost indestructible by tanning it with a strong formaldehyde solution: at emergence time the adult insect opens the cocoon with a few drops of a 3 per cent caustic soda solution secreted by a special gland. Yet more remarkable are the *Brachinus*, small beetles called "bombardiers" because of the detonation heard when they are picked up. The mechanism involved has just been discovered; it sounds crazy: one of the beetle's glands secretes hydrogen peroxide at a high concentration, a terrifying liquid in which biological tissue dissolves like sugar in warm tea; another gland secretes and mixes in a peroxydase, which brings about an explosion of the hydrogen peroxide. This is much the same thing as one of the methods suggested for the propulsion of space rockets. Other insects secrete caustic products similar to ordinary phenol.

The red ants are well up in the corrosive chemicals list, since, as I have already said, their formic acid solution is some 50 per cent. But they are not the only chemists in the family. Three classes of substances must be noted, poisons, attractants and repellents. In order to understand the use and distribution of the first and last of these substances we must go back a little to the classification of ants. For instance, one might expect poisons to be injected by means of a sting. But this organ is not universal in ants; it is found in the most primitive families, such as the *Ponerinae* and *Myrmecinae*, who hunt alone and are strictly carnivorous. The migrating ants, or *Dorylinae*, have a sting but hunt in packs. *Myrmica*, yet more developed, stings its victims, but is not exclusively carnivorous. Finally the more advanced ants, *Dolichoderinae* and *Formicinae*, do not have a sting, replacing it advantageously with various other mechanisms, of which the spraying of formic acid is by no means the least effective; and it is in these last families that behaviour is the most complex.

Now, poisons, repellent or attractive, are all secreted by a system of glands, fundamentally the same in all ant families; but what is interesting is the "modulation of the secretion" as it were, such or such a gland dominating according to species, and the others functioning very little or not at all. Thus in all ants we find: 1. Mandibular glands, found at the base of the mandibles, which generally secrete an alarm substance, awaking interest or aggression among the other workers. 2. Maxillary glands, associated with the mouth parts: their function is not properly known. 3. Pharyngeal glands, opening into the pharynx and associated with digestion. 4. Labial glands, situated in the thorax and producing a kind of saliva.

Of all these glands the mandibular are those that interest us more particularly at this point; but they are not the only ones to secrete substances influencing behaviour. Others are distributed throughout the body; for instance, the metasternal glands, opening on to the metathorax; it is said these produce the "nest odour", but the evidence is not very good and it would be safer to say we do not know what their function is. The poison glands are made of two filaments, floating in the general body cavity, which open into a vesicle at the base of the sting. At their side the Dufour gland is found, an azygous tube attached to the poison apparatus: it used to be thought that its function was to lubricate the sting but its role is more complex, for at least in one case (the fire ant) it secretes the track-marking odour. There are also the dorsal abdominal glands, opening on the sixth and seventh abdominal segments. They have been compared to the Nassanof gland of bees, which has so strong an attraction for the workers; but it must be admitted that we do not know their purpose in ants. Finally, we have the special anal glands, only found among the more developed groups and quite lacking in the primitive ones.

Moreover, it is only in the most primitive families that the poison and wounding apparatus follows the classic pattern (poison gland + Dufour gland). As soon as the diet is no longer exclusively carnivorous and ants start to take honeydew from aphids, then the Dufour gland greatly enlarges and starts to be used for marking the track. In another family, *Dolichoderinae*, the poison and Dufour glands are quite eclipsed by the anal glands, having nothing in common with the stinging mechanism; the former glands are used only to give alarm and for defence, and another gland is used to mark the tracks, the Pavan gland, at the end of the abdomen. Finally, with the *Formicinae* everything is again different, but in still another way;

there is no sting, but a poison sac (occupying a fifth of the abdominal cavity) full of a formic acid solution at a high concentration; the Dufour gland is still there and possibly plays a part in mutual recognition. The anal gland has not been found.

As to the chemical nature of ant poisons, the oldest known is that of the red ants, which at one time was the sole industrial source of formic acid. As we have seen, it was confused with vinegar: distillation enabled an appreciable quantity to be obtained from ants and this, no doubt, is the reason why it is the oldest insect venom to have been studied: it is mentioned in the *Philosophical Transactions* (London) from 1671 onwards. . . . This compound is highly insecticidal and poisonous even for the red ants themselves. As to the injected poisons, used with a functioning sting, less is known about them, due to the scarcity of the substance itself. Cavill and Robertson tried to get some from the huge Australian *Myrmecia*, which reaches a length of 2 cm., a very active and intractable insect. Not without receiving numerous painful stings, they tried stimulating the ants by means of electric shocks, which is the method used to get industrial amounts of bee venom: the shocked bees sting the substratum made of several cellulose layers, from which the venom can then be pressed. But the *Myrmecia* could not be persuaded to do this, and several hundreds of glands had to be dissected out, one by one. It is a so-called proteidic poison, that is it contains proteins. It appears that ant poisons are generally of this nature, although the *Myrmecia* venom is the only one to have been studied in depth. *Myrmecia* venom, like that of bees, contains histamine; hyalurodinases facilitating the spread of toxic compounds in the host tissues are also found. One of these last compounds is none other than a haemolytic protein resembling the melittin of bee venom, whilst the hyalurodinase recalls the phospholipases also found there. One might therefore conclude, without going too far, that *Myrmecia* venom is closely allied to bee venom. Nevertheless, an apparently very different poison is found in the fire ant; it contains a haemolytic enzyme which has insecticidal properties; but no more is known of it, except that the sting is painful and that it is often accompanied by necrosis of the tissues.

It is not always easy to distinguish between venoms and repellent substances: formic acid, for instance, is certainly both.

It must be admitted that the principal difference is that the repellent is sprayed on the assailant and not injected. Another good example is the substance iridomyrmecin, isolated by Pavan from the Argentine ant, *Iridomyrmex humilis*: this is the first ant chemical that has been fully identified and subsequently synthesized. I will not inflict a course of chemistry lectures on my readers; it is sufficient to say that iridomyrmecin has nothing in common with formic acid. The Argentine ant's anal glands produce it. Many other substances with strange and sonorous names have been identified from ants: hexanal, heptone, tridecanone, iridolactone, terpenes and terpenoids: all this is due to a miraculous instrument, the argon chromatograph, which takes in a tiny bit of the substance at one end and at the other draws a handsome curve giving the chemical constituents and the amount of each substance present in the mixture. The alarm substance emitted by the mandibular glands contains dendrolasine, citral, citronellal, etc., all highly aromatic substances. The chemical composition of the track-marking substances is not known.

In addition we must note that within any species the different castes can secrete highly aromatic substances very different from each other. It is said that the corpses of the sexuals smell of roses whilst those of the workers give off a disagreeable smell of excrement. There is a species of *Pheidole* in which the soldiers have a horrible smell whilst the workers only secrete a trackmarker, which does not offend the nostrils at all. During the nuptial flight in certain species of *Lasius* the males give off a mixture of terpenes and a compound on an indole base. This very probably serves directly to attract the females, or it may leave a scented track in the air, along which the females can fly. Each species of *Lasius* produces a different mixture, the proportions of which have been carefully established by analysis.

CHAPTER VII

THE BRAIN AND SENSES OF ANTS

Can they learn?

The ant brain

NOT MUCH IS known about the ant brain, one might as well say nothing. On the whole it does not seem to differ much from that of other Hymenoptera; but so little experimental work has been done, apart from a few anatomical studies. It must be noted, for instance, that surgical intervention is very difficult in ants, both because of their small size and because of their social life: any ant that is wounded or whose behaviour is the least abnormal runs a considerable risk of being sacrificed. The *total volume* does not tell us much: we know it is related to the development of the mind—or, rather, not the actual volume but the *ratio of brain weight to body weight*, for the elephant's brain is actually bigger than a man's. Now with the red ants the brain

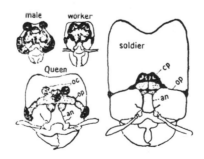

FIG. 34

Heads of different *Pheidole instabilis* castes (the ocelli indicated by black discs) showing relative size of the brain. All drawn to the same scale (after Wheeler).

is about the same size in all three castes; but then their size does not differ very much. With markedly polymorphic species, such as *Pheidole instabilis* (see Fig. 34), it is found that in spite of the enormous difference of size between castes, the brain is about the same size throughout, which means that, when considered in conjunction with the body weight, the worker has, proportionally, a *much bigger brain*. Now, it is a fact that its behaviour is much more complex. The worker's predominance in this respect is yet greater if no account is taken of the optic lobes, which are only concerned with the eyes, much reduced in the worker.

Where is the "seat of intelligence"?

Great importance has been given to the *mushroom bodies*, nodules of this shape, peculiar to insects. Forel had already noted that their size in ants is in no way related to the size of the eyes or muscles; they are always small in males, big in workers and intermediate in the queens, which led Forel to suppose they were the "main brain" of the ant. But Wheeler observed that though the mushroom bodies are in fact well-developed in certain species, this is by no means the case in all, far from it. In addition it is not the *major* workers that have them most developed, but the medium-sized ones: in the big workers they are often no bigger than those of the queens. Nevertheless, and this supports Forel's theory, the most surprising instincts are mostly found not among the biggest workers, but in the small and medium-sized ones.

The senses of ants

For some time now considerable progress has been made generally in the study of the senses of insects. It all really springs from the wonderful work of von Frisch. This learned German is a genius comparable to Pasteur and Ampère. Frenchmen, now grown, will remember at one time in school Pasteur's experiments were explained to them. By the use of a few glass globes, cotton wool and some bouillon, the inexistence of spontaneous generation and the existence of microbes were proved. Alas! we were much too young to understand most of it. For, blindly following an absurd theory of education in abstractions, we reserve the beauties of experimental science for whipper-snappers of ten years old and class it as the least important subject. But to make up for it we lose no opportunity of stuffing their minds with a hotchpotch of mathematics, which 70 per cent of them will never use. It is a superstition going back at least to ancient Greece: only mathematics, with their enforced detachment from reality, are deemed capable of forming the minds of the young, whilst reasoning from the results of an experiment does not have, it seems, the same educational value.

Please forgive me this digression; but when for forty years you continually hear this same nonsense, and see it carried out, there is some reason for disquiet. Let us, rather, consider Ampère, another famous case, who one day heard talk of Œrstedt's experiments with the magnetic needle, which an electric current can cause to deviate. He shut himself in his room for a week and, with the help of a few bits of copper wire, a compass needle and two or three batteries, he carried out on his table the admirably simple experiments on which modern electrical science is founded. Ampère's table is famous and it is still to be found, together with some of his admirable contrivances, at the *Conservatoire des Arts et Métiers*.

Well, von Frisch is of that stamp. This little man, deaf and silent, has but one love, bees, and for forty years he has studied them: he has no need of complicated apparatus (and that, moreover, is an attribute of the greatest men: one always admires both the elegance of their experiments and the simplicity of their equipment). A hive, a few bits of wood and a chronometer suffice him. And out of this came not only the wonderful discovery of the bees' language—which I have told in another book—but also, which is much less well known, the most complete study there is on the senses of these insects: sight, smell, taste, touch and balance. The methods are always the same, whether we are dealing with bees or any other insect: it is worth while lingering over this a little. Let us see how the eyes are constructed.

Ants' eyes and sight

Eyes are very unequally developed, but are much smaller in the majority of cases than in wasps or bees. Only a few primitive families (*Myrmecia*) have large eyes; on the other hand many species are blind or have only very reduced eyes; for instance, the robber ant, *Solenopsis fugax*, who digs its galleries in the thickness of the red ants' nest walls. Ants, more generally the sexuals, in addition to compound eyes made of many facets, or *ommatidia*, also have simple eyes or *ocelli*, with very few elements, which are generally situated above the compound eyes. They are but rarely found in workers. It is said that they are used above all for night vision, because in a large African nocturnal ant, *Aphaenogaster gemellata*, the ocelli are enormous; on the other hand they are small in *A. gaster testaceopilosa* and *A. depilis*, which are diurnal. But one example does not constitute proof. The question of the use of ocelli in insects in general, and not only in ants, is, moreover, one of the most complicated there is. Let us get back to physiology.

For instance, colour vision: how do we know whether an insect sees colour or not? It is by no means enough, as some old writers supposed, to see an insect try to get honey from a flower in the design of a tapestry, and to suppose from that that it can distinguish the colour: possibly the shape interests it, or some contrast with the background colour, which would exist even if the creature saw the universe in terms of a black-and-white photograph. Von Frisch then conceived the idea of having a table covered with squares of all shades of grey, randomly distributed; among them was one blue square, if blue was the colour he was testing. The bees were attracted to the blue square by putting a few drops of sugar syrup on it; then when they had got used to sucking up syrup there he moved the blue square and put it down somewhere else among the grey squares. Now, if bees saw the world merely in black and white there is no doubt that they would inevitably confuse the blue with a grey of equal luminosity; but if they see colours as colours, the moving of the blue square will not trouble them, and they will fly straight to that: this last is, in fact, what is observed in the case of bees.

In ants the grey-square method has given uncertain results; workers have preferred the optocinetic reflex system. It is a strange reaction, peculiar to invertebrates. In the higher animals, including man, it is far from being as clear-cut: if you put animals on a platform surrounded by a cylinder whose walls bear vertical stripes, the creatures tend either to "move on the spot" facing the direction in which the bands are made to revolve or, sometimes, to move in the opposite direction. Obviously this apparatus can be used to study acuteness of vision, for instance, by gradually decreasing the width of the bands, a point is reached where the subject no longer moves as the cylinder turns: it no longer sees anything. One can also cover the cylinder with bands of all the different shades of grey alternating with bands of the colour to be tested; then if they only see the world in black and white, there will be a grey they are unable to distinguish from the test colour, and they will not react to the cylinder's movements. Moller Racke made this experiment; but only on one species, *Lasius niger*, which does not seem to be able to distinguish colours from greys. From which several writers have concluded—too hastily in my opinion—that "the ants" cannot see colours. To begin with colour vision is very common in insects, and *Lasius* might just be an exception. Secondly, big differences between neighbouring species can be found. For instance, in mammals colour vision is not very common except in primates and man; rats and mice do not see colours, but the squirrel does, though it is close to the former in the zoological scale. However, it must be pointed out that ants, it seems, have no use for colours, whereas they are most useful in enabling bees to find flowers.

Perception of form

We have better information on ants' visual powers in regard to shape. Once again we draw on the German school for our information. Jander and Voss studied the red ant. In the middle of a flat, uniform enclosure they put a few pupae which the ants were expected to carry back to the dark and humid nest on the boundary. The nest was marked with cards bearing different patterns. Round the border of the enclosure a number of other cards with more or less similar patterns were placed. The ant then had to recognize the right pattern: or rather to re-find the correct card after it had once found the nest by chance. It was found that there was a well-marked recognition of certain shapes, but that this was upset by a number of innate preferences, which introduced additional errors. For instance, *ants definitely prefer vertical stripes to horizontal ones*, and simpler shapes to complicated ones. This last preference is unexpected, for it seems to be unique among insects, who generally prefer the more complicated and split-up patterns. At first sight it is not easy to see why ants act to the contrary. The ants can, with further training, be taught to overcome their preference for the simpler figures and be made to choose the more complicated ones; and here again they differ from bees. *Bees prefer the most split-up patterns and no amount of training will get them to change their preference.* On the other hand training will not make ants change their predilection for vertical stripes: it is still fairly strong even if the "good" signal is a card carrying horizontal stripes. The opposite obtains with bees, which also prefer vertical to horizontal stripes, but can be trained to reverse their preference. The reason for these innate preferences and their unequal persistence is more than I can explain. . . .

But Voss has gone further than that, showing a very German ingenuity and attention to detail. Let us still consider the matter of the vertical stripes the ants like so much. They quickly lose their powers of attraction if they are inclined more than 10° from the vertical; on the other hand they must be sufficiently tall. They scarcely interest the ants at all if they subtend an angle of less than 26° above the horizon. In addition the two upper and lower halves of a band must be continuous and exactly so, or their attraction literally disappears.

How is this to be interpreted? We do not know. Obviously, it has been supposed that this predilection for vertical bands springs from the fact that the red ants are greatly interested in trees and their trunks, because it is there they find their cherished aphids. But some tree trunks slope and others bifurcate; how is it then that such slight deviations of the bands from the vertical cause the ants to lose interest?

On the other hand, in spite of their preference for vertical lines, ants can nevertheless distinguish horizontal from vertical lines but *not lines inclined in one direction from lines inclined in the other*; strange to relate the ability to discriminate in this way is acquired only with difficulty in the animal kingdom, by animals as far from the ant as rats and octopuses.

Vowles repeated Voss and Jander's experiments, using a slightly different apparatus: he used a T-shaped apparatus from which the insects had to find the way out, which was in one of the arms of the T; a pattern was put just above the exit.

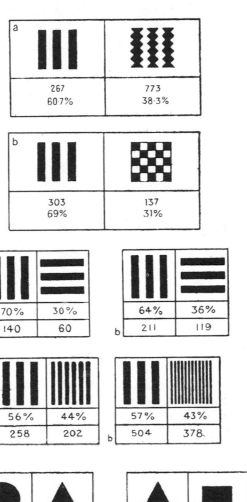

FIG. 35

Preferences of ants for certain shapes. Upper figure: number of ants in experiment. Lower figure: percentage choice. Note that in the two upper diagrams there is a preference for a simple design (the reverse of bees). In the four middle diagrams there is a marked preference for simple vertical lines. The bottom two diagrams shows that ants distinguish between a circle, triangle and square, which is not found in bees (after Voss).

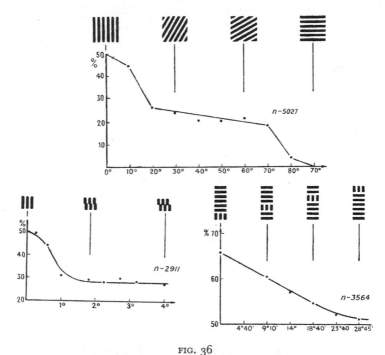

FIG. 36

Upper curve: slanting the stripes lowers the number of spontaneous choices. Abscissa, angle of slope of stripes. *Lower left:* a slight staggering of the stripes greatly lowers the number of spontaneous choices. *Lower right:* three vertical stripes are shown to the ants at different heights above the horizon. Abscissa, angles above the horizon. Only stripes just above the horizon are chosen (after Voss).

The same results were obtained; with the singular observation *that different nests showed very differing skills in learning* and manifested very dissimilar behaviour patterns.

The visual acuity of ants can be measured by calculating the angle of opening of the visual organ or *ommatidia*. When trials are made, progressively decreasing the width of the test bands, it is seen that it measures 34 minutes of a degree, which is much finer than their anatomy would lead us to suppose. This has been successfully explained, not only in ants but also in other species of insects. But it is not easy to expatiate here because one would have to bring in complex theories of cybernetics. It is sufficient to note that certain physiological arrangements, such as the nerve connections of each ommatidia, have the effect of increasing the contrast images make against their background, thus augmenting the sensitivity and power of ant vision. In a word, one can conclude that ants, though very short-sighted, like all insects, can see fairly well on the whole.

Direction-finding

This is particularly noticeable in their *orientation in the open*. It is certain that they use the sun as a reference point, and allow for its movement, as do bees. It can be proved by repeating all the experiments made with bees, which it would be tedious to describe here. In particular, if you show a *Myrmica* a reflection of the sun in a mirror it becomes completely confused, but it is an ant relatively low on the ladder of ant psychism. The same experiment either fails or works badly with red ants. As with bees, ants *do not rely on the sun alone to find their way*; if it is covered they, like their honey-gathering neighbours, use patches of blue sky still to be seen. For, again like bees, *they can distinguish between polarized and ordinary light*. And each patch of blue sky has a particular appearance when seen by polarized light, according to its position relative to the sun; thus, starting with the first one can reach the second. Note also that if you make some ants follow a particular path in ordinary light and then cover part of the track with a polaroid (a plastic material that polarizes light passing through it) the ants become confused when the polaroid is turned, thus altering the plane of the light passing through it. As with bees once more, ants can even distinguish the sun when the sky is completely overcast, provided that the clouds are not too thick. In this case bees use the image of the sun made by the far ultra-violet light which can penetrate cloud; von Frisch proved this in spite of the opposition of more than one physicist, who would not admit the possibility. Ants also seem to use the lighter area formed by the sun in white clouds. In the laboratory Jander has shown that they can orientate as well on a luminous area of diffused light as on a concentrated source of light.

Finally, once more as with bees, ants use *physical objects*, a large pine here, a rock there and a bush elsewhere. . . . They can find their way (*einkalkulieren*, as Jander puts it) using at the same time a black panel and the sun or the polarized image of a piece of blue sky: they say (so to speak), "We must go between the black panel and the sun up to five o'clock, but always in front of the black streak after, even though the sun is not in the same place", and if you move the panel they are greatly disturbed.

In fact the numerous animals, from insects to fishes, who "steer" by the sun, *make allowances for the movement of the orb*: this presupposes both *a very exact awareness of time* and a very complicated mechanism of internal integration. This has been very closely demonstrated in the bee by one of von Frisch's most brilliant pupils, Renner. For example, an ant (or a bee), shut up for a while in a dark box, takes up its original route when released and makes allowance for the movement the sun made during the insect's imprisonment. But this does not happen, or only rarely, with ants coming out of hibernation. *It seems probable that this means of calculating movement must be relearnt after hibernation*, and that ants have no innate knowledge of, but only the latent power of learning this behaviour. In the same way guidance by means of physical objects in the neighbourhood presupposes *a good memory, in fact an exceptional one for an insect.* They have forgotten nothing after a week; and remember not only an isolated object, but also complex combinations of reference points. I myself have seen this in the course of involved maze experiments (see p. 185). And if one can believe Wellenstein, ants' memories go well beyond that: in the autumn he marked a good number of workers with paint and found them following the same routes in the ensuing spring. If this is the case ants are easily the memory champions of the insect world.

Finally, there is another reaction to light bearing on orientation: this is called *phototaxy*, that is movement towards light (positive phototaxy) or from light (negative phototaxy). Red ants are photopositive when they go out to forage and photonegative when they return to the nest.

Sense of balance

We have behind the ear a curious formation made of three semi-circular canals, corresponding to the three dimensions of space. The way these organs, and a number of other similar ones found in the animal kingdom, work is always the same: in the liquid in the canals a number of mineral particles, of about

THE BRAIN AND SENSES OF ANTS

the same density as the liquid, float; the least movement of the head or body affects the particles, which, owing to their inertia, are slightly retarded in following the movement; they thus brush against a number of sensitive hairs in the canals. The central nervous system is then informed, according to the position of the hairs affected, of the movements of the head or body.

But insects have no similar apparatus. Nevertheless, there is no doubt that they can keep their balance and, at times, can fall on their feet with the agility of a cat. And people have asked themselves how this could be, until the celebrated Professor Lindauer, the great bee specialist and a pupil of von Frisch, some even say his equal, discovered the key to the mystery: bees have special hairy patches on the joints of the petiole, the neck, legs, etc. Each is very close to the immediately contiguous segment and thus rubs against it whenever a movement is made, however slight. These hairs also send messages to the central co-ordinating area of the brain where the information is combined if need be with information coming from other areas. An order is then sent to the motor system which can lead to a correction of the body position if necessary.

These balancing organs have been found in other insects and Markl, in particular, has found them in red ants. He blocked some of these organs by covering the hairy patches with wax and was thus able to see which were more important: in fact, he found a kind of scale of importance. In the maintenance of balance the following was the order of descending importance: the receiving apparatuses of the neck, petiole, antennae and coxa: those of the gaster did not play an important part and an ant without an abdomen can maintain its balance. The hairy patches also play another part in addition to that of maintaining equilibrium. They inform the nervous system about the body's position in the horizontal plane. It is only when all these organs transmit a similar message to the central nervous system at about the same time that it is interpreted as being an indication regarding weight rather than the actual position of the body.

Combination of visual and gravitational orientation: "Geomenotaxy"

But the possibilities of "recombining" sensory messages go further still, and the "control co-ordinator" can combine information from two different sources, sight and the sense of weight.

Here I must go into a little more detail and refer once more to that most studied insect, the bee.

One of von Frisch's most important discoveries was the bees' dance, as everyone now knows. On its return to the hive the honey-gatherer who has found a new source of food performs a dance in the form of a flat figure 8: the angle that the long axis of the 8 makes with the vertical (in other words with the direction of gravity) is equal to another angle, that made by two lines, one from the hive to the sun, and the other from the hive to the newly discovered food source. In other words it is a question of transposing two messages absorbed by the sight receptors (the respective directions of the sun and food) to new coordinates using gravity as the reference. And one can easily follow the progress of the change by slowly tilting the vertical comb on which the bee dances to the horizontal. Then the long axis of the 8 stops being based on weight and points directly to the food, like the needle of a compass. If on the other hand you move the comb back again, the relative angle with the direction of gravity is completely restored as the comb reaches the vertical.*

Obviously such an operation (which, it must be understood, is completely automatic) far exceeds anything one thought possible from the nervous system of insects. Naturally, one asks whether it is only bees who possess these strange powers, and whether other insects can do likewise, particularly social insects, whose degree of awareness is nearest to that of bees. And one finds that ants can, more or less, do the same thing. The mechanics of the operation are not as perfect as in bees, no doubt for the good reason that they have no need of it. Their system of news-spreading, when a promising source of food has been found, seems quite different, in many cases it is still not very well understood. Markl introduced some ants to a vertical disc and faced them in a certain direction in relation to a light source, whose rays were parallel to the disc; they found their way back to their nest, for instance. Then he put the light out and noted whether the ants, thanks to their memory of the gravitational pull, could still keep their sense of direction. In effect the workers were able to do so, but with certain characteristic deviations ("errors"). For example, if the angle between

* See the author's *Animal Societies*, Gollancz, 1968.

the vertical and the goal is between 0° and 90°, the ants tend to go too high; if the angle is more than 96° they tend to go too low. (This resembles certain errors made by bees in similar transposition experiments.) These errors must come from some defect or peculiarity of the central co-ordinating areas, because Markl found that blocking several of the hairy patches sensitive to weight made no difference. Neither the inclination of the disc from the vertical, nor the putting of small metal weights on the ants' gasters affected these errors.

Chemotactic senses

We do not have many facts about the senses of smell and taste in ants. Nevertheless, an old work of Schmidt's informs us that their sensitivity to different sugars varies according to species. The carnivorous species, such as *Myrmica*, only react to seven kinds of sugar, whereas those with a more varied diet including honeydew will take a dozen sugars. The taste organ is situated in the antennae and they do not seem to have another on the tarsus as have bees (at least in the case of the *Myrmica* ants). The "sweetest sugar", that is to say the sugar recognized by ants as sugar at the lowest dilution, seems to be saccharose (as is also the case with all insects). Certain complex sugars, such as melezitose and raffinose, which can be found in honeydew, also appear to taste very sweet to them. But all this is very old and the research should be continued on other species. More recent work has been done on the chemical stimuli that induce ants to carry out the bodies of their dead companions: such substances are related to decomposition products from organic tissues. Finally, Vowles managed to perform microsurgery on the brains of ants; the rapid conditioning of ants to a smell is cancelled by the excision of certain definite points in the brain.

Hearing

We do not know whether ants hear anything at all, and the point is equally uncertain with reference to bees. Examples of reactions to sound that have been put forward here and there probably arise from a *sensitivity to vibrations*, which is highly developed in insects. Some grasshoppers, otherwise completely deaf, react to sound because they perceive the vibrations it causes by resonance on the substratum.

The *production of sound* is well known in bees and ants. These two insects continually emit not only a buzzing of varied tones, as does the bee, but also bursts of ultrasonic sound. In 1874 Landois found a sound-producing organ at the base of the abdomen of a *Ponera*, and a few years later (1877) Lubbock described a similar structure in *Lasius flavus*. These organs are present in many families and consist, in their simplest form, of very thin, parallel and transverse *striae* extended over a small area at the base of and above the gaster's first segment: it is covered by a prolongation of the previous segment, the edge of which is chamfered so that it can rub against the *striae*. Sound production in ants has not been much studied. Raignier remarked that it was not all produced in the ultrasonic band, and that sometimes one could hear a slight stridulation from *Ponerinae* and *Dolichoderinae*. *Megaponera* stridulates in chorus when one removes the leader, and the ant turns back to the nest, or when an expedition is about to set out. It is said that Nigerian children know this well and claim that they can drive the ants mad by whistling in a certain way (it is true that the stridulation of these enormous ants is something like whistling). The workers of *Myrmica schenki* also stridulate when attacked by other insects. This leads to great excitement among their companions in the immediate neighbourhood, but they do not necessarily go towards the attacked ant to help it (thus behaving like man!); it is simply that the general excitement increases the random journeys and hence the likelihood of encountering aggressors. Nachtwey has described a wide variety of stridulating mechanisms, which he calls "phonotors", in a great many ants. These mechanisms are very complicated, having resonating chambers, vibrating stalks, etc., equalling or surpassing anything of this kind previously known in insects. But he only carried out a simple anatomical and descriptive study; that is to say we do not have the least idea what the sounds they make are like.

In all stridulating insects the possession of sound-emitters goes with the presence of receivers capable of registering the sound. Ants are no exception; *their "ears" are found in the anterior tibias* and are much like those of crickets. In *Camponotus* the organ in question contains 18 to 20 sensory fibres and reacts to soil vibrations of from a hundred to three thousand cycles per second.

Are these signals noticed? Are they not, perhaps, just a background noise with no behavioural significance? It is very difficult to know. Recent research work has cast some light on the matter. Markl, studying the fungus-growing ants, *Atta*, noticed that they stridulated when shut into a vessel from which they could not escape. Much of the sound is inaudible to man, as it runs from 20 to more than 100 kilocycles per second. With *Atta* it seems that there is a particular difficulty to be dealt with when studying stridulation, because these ants have a supplementary alarm system, a mandibular gland that puts out a highly aromatic alarm substance, citral. Markl overcame this drawback quite simply by cutting off the heads of the *Atta* ants, which does not stop them stridulating. What vitality insects have! But the sound does not attract ants *unless it is carried by the soil* (which shows that it is a matter of sensitivity to vibrations) and it sets off a strange activity—burrowing into the soil where the transmission is greatest. It is only the *minima* workers who burrow. The queen can also stridulate, and much louder too as she is very much bigger; perhaps this gives her a better chance of survival if a tunnel caves in over her. Also, who knows if the normal function of this stridulation is not to guide the worker ants when they are digging the complex galleries of the nest?

In another fungus-growing ant, *Pogonomyrmex*, it is possible that more involved messages are sent, because the nature of the stridulating mechanism is more complex. In this case too the stridulation is produced by the rubbing of the mid-dorsal area of the gaster against a series of *striae* found on the posterior petiole; it is the successive rising and falling of the gaster that produces the sound. Its frequency is lower than in the case of *Atta*, consequently the sound travels further, though its intensity is less; once more it is the substratum that transmits the noise. The ant can vary three factors: the overall intensity of sound (louder when the gaster is raised, less when it is lowered), intensity during the rising phase (which can vary independently), and finally each stridulation is divided into successive emissions separated by pauses of variable duration, all controlled by the movement of the gaster. Thus a possible apparatus for transmitting information exists, but it has not been shown whether the ants use it effectively.

Sense of time

Bees were easily trained by Renner and others to come for sugar at a certain place and time. This is understandable as they collect pollen and nectar, which are often only produced at certain times and for short periods. It is not the same with ants. The aphids' honeydew is exuded whenever they are stimulated by the workers, whatever the time. Hunting continues day and night, provided the temperature is high enough. There is thus no apparent reason why ants should have a "time sense". And, in effect, modern experiments have shown that up to the present ants have never been trained to come for food at any given moment. Their time sense is shown in another way.

Activity rhythms

There are diurnal ants and strictly nocturnal ones. According to Bernard fifty-four of the fifty-eight Sahara species are nocturnal, which is explained by the insupportable heat of the Sahara soil during the day. Nevertheless, species travelling by day, such as *Cataglyphis bombycina* and the famous reservoir ant, *Myrmecocystus*, are known. Some of these ants are covered with silver-coloured hairs, so that one might say they were aluminiumized! Obviously it is possible that this arrangement may reflect a considerable part of the incident radiation. But a problem still remains in view of the microclimate laws recently postulated by bio-climatologists; the heat at ground level on a summer day is much higher than that at the level of the human face, where our weather instrument screens are usually placed. At ground level on a summer day in the Sahara temperatures over 70°C have been recorded: one wonders how an insect can move about in these conditions without dying of heat. There is a strange physiological problem here which seems to me to have been neglected.

Then again, ants in our climate, such as our red ants, which go out during the day, take a rest when the temperature is too high, although they readily forage during the hot nights. A whole series of questions arises. I have shown that, in effect, they bring back about as much game from these nocturnal expeditions as from the daytime ones. Now they are supposed to hunt by sight: at night then they must use other senses (see p. 59).

In many insects the rhythm of activity, be it diurnal or

nocturnal, depends on an "internal clock", regulated by their metabolism. This can be proved by shutting some insects in either a dark box or one continuously lit; under these conditions the various activities are carried out at the usual times. At least that is what happens with *Veromessor andrei*, but not with *Formica polyctena* whose rhythm (in any case not strongly marked) soon disappears under these conditions.

Nevertheless, after worker ants have been in a dark box they set off without fail towards their nest as soon as they are set at liberty: undoubtedly they are steering by the sun. This means they must be allowing for its apparent movement, as bees do under similar conditions. This implies that they do have a certain time sense, but one used only for orientation and not for foraging. Dobrzanski pointed out that the ants' food is always available in nature, whereas the bees' nectar is only obtainable at certain hours.

Can ants learn?

In spite of the large number of naturalists interested in ants, not many experiments have been made on behaviour. I believe watching ants so enthrals the observer that he never even thinks of interfering. But, as we have already seen, are going to see and shall see again, experimentation is the mother of science: it alone allows one to interpret an observation and to test it to see if the conclusions drawn are sound.

When one sees *Formica* carry out so many actions (as a whole, not individually) appropriate to the desired end one naturally puts the same question as one would in dealing with a higher animal, a mammal, for instance: are they able to do all that instinctively from birth or do they have to learn? Perhaps the question is badly put, but intentionally; no doubt it implies, as we shall see later, the subtle suggestion and error of comparing an ant, even remotely, with a mammal. This is a problem we have tried to solve in various ways, and the American biologist Schneirla was one of the first to start a series of experiments with this object.

He used an apparatus dear to experimental animal psychologists, the maze. But for you to understand the reason for the choice, and the utility of this instrument, I must be allowed a digression.

The maze and the white rat

The white rat, that long-suffering laboratory animal, was first put in a maze many years ago, with the object of studying its behaviour. At first the maze was constructed anyhow, but was always very complicated. One of the first "labyrinthologists", Small, took the green hedge maze of Hampton Court, in England, as his model. Hungry rats were put in at the entrance and they had no chance of getting any food until they managed to thread the maze and reach the centre. The strange thing is that they managed to do so as well, at times even better, than man: and by that I mean a man prevented from using human aids, such as Ariadne's thread or a map. (The humans tested had their eyes bandaged and had to follow the paths of a small maze with a stylus.) It can easily be seen that this apparatus can be made into a kind of microscope to be focused on this or that aspect of behaviour. For instance, one aspect was at first confused with intelligence, but the reality is much more complex: it is the power of learning even more complicated mazes. It was thought that a scale of psychism could be established in this way. From this angle one can find very big differences between species and even between individuals of the same species, but the results are difficult to interpret.

If, after the subjects have obtained a certain *score* they are left to rest a certain time and then tested again, it is *memory* that is being studied. One maze having been learnt, the subjects can be made to learn another, similar or different: one is then studying the possibility of *the general application or adaptation of a solution*. We might also try opening a short cut, thus enabling part of the run to be cut out; it is then possible to see if the subjects have sufficient flexibility to adapt this new solution to an already known problem. Other investigators have modified the lighting, positioning, colour of the walls, or have put the maze in the dark: or they have operated on the subject itself, removing one or more of its sense organs. In this last case orientation mechanisms are the subject of study. The technique has aroused so much enthusiasm that more than eight thousand papers have been published on rats in a maze. Which is far too many. One should be interested in the problem, not the apparatus considered as an end in itself; no procedure can solve all problems, and this one is not without defects. But this is not the place to go deeply into that.

Schneirla had the idea of comparing the behaviour in a maze of ants and rats; he used a maze the size of which could be adapted either to an ant or to a rat. The ants (*Formica lugubris*, a fairly near neighbour of *polyctena*) were introduced one by one, which, as we shall see later, was perhaps a serious mistake in making such experiments.

First of all it was found that both animals, rat and ant, could master the maze, the first much more quickly than the second: the rat needed some fifteen attempts and the ant more than thirty. Moreover, the ant went through a series of stages which were "telescoped" in the rat. In the first stage the ant wanders around at random until about the eighth attempt. She then tries all the cul-de-sacs with desperate regularity: little by little she is seen to stop about half way along them; then she only goes a third of the distance; finally, she stops a little after entering the blind alley. Basically the behaviour of the ant resembles that of the "maze machines", crude robots constructed about 1930, before the cybernetics era. Fitted with a photo-electric cell and simple instruction mechanisms on a trial-and-error basis (each success increased the chances of entering the same alley, each failure decreased it) the robots learnt an easy maze and enabled the theoretician to prove how well-found were his machine and his theory of rat behaviour.

The irony of the situation was that the rat did not behave like that at all! In order not to see it, the research workers must have been completely obsessed by a frantic desire to be right, in face of all the facts; the curse of experimental science. The ant, on the other hand, does resemble the robot. In place of gradually registering the blind alleys, rats are mostly content to hesitate so briefly that it is difficult to see whether they pause or not. It is clear that from the first run they have integrated all factors into a coherent whole and have understood the situation almost from the start of the trial.

Now, it is just this *integration of information into a whole that seems so difficult for the ant*. Towards the fifteenth attempt, when slowly and painfully a good number of blind alleys have been eliminated and it is only a question of retaining the information, organizing it, signs of trouble appear, such as rapid circling and

stops followed by quick runs. It can happen that, having learnt
to avoid a new cul-de-sac, the ant temporarily forgets the
previous one, the avoidance of which had been perfectly
acquired. But catastrophe occurs when one forces the ants to
return from the end of the maze to its start: this rats do very
well and the "reversal" does not seem to worry them. As to the
ants, alas! they behave as if they were fresh subjects, "naïve" as
the Americans say, starting from scratch to learn an unknown
maze.

However, there is one case in which ants have the advantage.
If after a series of turns in one direction they are made to turn in
the other, ants learn the inverted turns neither better nor worse
than usual. Rats, on the other hand, especially if they are made
to alternate the inverted turns with others in the "easy" direction (the one they had previously got used to), gradually begin
to show signs of acute distress: they make standing jumps, throw
themselves against the walls of the maze even violently enough
to get bloody noses. Later the trauma induced by this alter-
nation can become so strong that they stop, as Schneirla says,
"as if blocked by an invisible wall". Faced with this test the
robot-ant is less worried, although it shows some slight degree of
disturbance: at the goal, after each test, it sucks up a little less
of the syrup placed there as its reward.

All this resembles very closely the behaviour of many other
insects in similar circumstances. With several pupils I remember
starting an intensive study of how *Blatella* (the common cockroach of our kitchens) behaved in a maze. We were then a
handful of workers without a halfpenny of funds; apart from the
fact that we loved insects and wanted to study them, an essential
factor in this choice was the fact that it did not cost much! The
years we wasted in that way! But on reflection perhaps
"wasted" is too strong a word, for the value of a piece of
research work is in no way related to the funds it needs to carry
it out. It is true in biology and even in physics. I have heard
more than one physicist deplore the fact that cyclotrons, costing
several hundred thousands of francs, were to some extent
killing physics; although it is always possible, with very little
apparatus, to renew this science in the way Ampère did on the
corner of a table, not so long ago. . . .

As to us, shut in a dark silent cellar with our cockroaches in two or three flasks, our maze made from a few bits of metal, and our notebooks, we found that it was nevertheless possible to work and to find some strange biological problems. Strength of imagination and "making-do" will succeed in the end; it is a hard lesson but a very useful one. We do not regret having learnt it. After months and years, after countless repetitions (some experiments needed fourteen thousand successive tests), a few conclusions began little by little to appear. One of them was precisely this stiff and automatic nature of the insect's behaviour, with improvements coming only by slow steps and with very little power of adaptation to different problems. Then, after considerable reflection, I wondered whether we were not on the wrong road, or at least whether we had not put the cart before the horse. One of my workers had in effect just elucidated the basic laws of *exploration activity* in cockroaches; that enigmatic and universal activity present in every animal, the essential basis of all its other behaviour, and particularly of behaviour in a maze.

I cannot expand on this theme, it is enough for the reader to know that when an animal has eaten and drunk enough, when its sexual appetite is at rest, then, instead of sleeping or doing nothing, as the old psychology would have us believe, *it ceaselessly explores the neighbourhood until all the details are registered in its memory.* A veritable "hunger for new experiences" takes possession of it. Slowly this idea of exploratory activity has forced animal psychologists to reconsider the basic ideas of the science.

We then realized that many facts we had noted may not have been due to learning but only to the development of the exploratory urge over time. And, in addition, that each kind of maze gave rise to a new kind of exploratory activity.

This may sound discouraging, because research so often seems to "advance backwards". But there is no need to be worried; all it means is simply that the ideas on which the experiments were based had not been sufficiently deeply studied. Such happenings are common in a science as young as animal psychology. And, in the course of this book, which is only half about ants, the other half being to show you how scientists work, we have seen many cases of such retrograde steps.

Nevertheless, we pushed our experiments with cockroaches further than Schneirla did his with ants, which is why I felt a

little uneasy when re-reading the works of this American biologist. If the study of an insect with a relatively simple biology,
such as a cockroach, can present so many traps for the unwary,
what could ants not do?

First, we must lay down some sensible rules. The first is that
the animal must be studied as far as possible in its natural
habitat, otherwise a series of disturbances of unknown extent
can be released. The old experimenters were not so careful;
we have become more critical. When Loeb showed that a caterpillar of *Porthesia* put in a glass tube moved towards the sun's
rays, whose heat would kill it, he thought he was demonstrating
the forced, automatic and inflexible nature of the light "trop-
ism". But in nature this caterpillar does not necessarily move
towards the sun, and if the solar rays bear down on it too heavily
the creature simply creeps into the shade. . . . It is all very
tiresome, but easily explained: the caterpillar is familiar with a
certain surface, the leaf, and it is put into a glass tube, a very
strange environment for it; then if, by chance, it should want to
get out of the sun, it can only do so by going backwards, a
difficult operation for caterpillars; but the tube is too narrow to
allow the creature to turn. So that finally what we get demonstrated is no more than an abnormal phenomenon. . . .

In my view, the case of Schneirla's ants is much worse. With
my cockroaches there was reason for some optimism as regards
habitat, for we were working in a cellar where very many cockroaches, escapees from our breeding vessels, had, disastrously,
established themselves; the habitat could be considered as all
too natural, alas! But ants are social insects; they scarcely ever
move about alone and their companions are never far away and
mostly in quasi-immediate contact with them: as we have just
seen, they are ceaselessly exchanging a mass of stimuli from
which their social behaviour arises. Now, first of all, Schneirla
put them into an experimental nest, where the number of
workers was very small, a situation far from natural, and then
he let them into the maze only one by one. Again, ants do not
just leave the nest by chance, they do so for a purpose; it may be
to collect twigs, or the sugary secretions of aphids, or to go hunting: they have, as the jargon has it, a "special motivation".
Now the research worker "rewards" the worker that has
threaded the maze with a piece of meat or sugar, without

enquiring whether in fact it was this the creature was seeking. Might this not well be a reason for the poor learning scores found in ants? What happens if you offer meat to a worker looking for sugar? Nobody quite knows. On the other hand we have the exploration behaviour—about which little is known except that in ants it is done in a singularly precise and careful manner—was this exploration carried out sufficiently by the test ants? As with cockroaches, *are we watching a learning process or an exploring one?*

Another thing: how is it that workers take so long to learn a maze that is quite simple, compared to the complex labyrinth of their own nest? The least one can say is that, in their natural habitat, the workers show none of that hesitation described by Schneirla: on the contrary they appear to be full of self-confidence, always "knowing what they have to do", even if to our eyes it appears to be senseless—but that is another story.

To solve the problem I think one must get out of the laboratory, and above all not copy too closely the experiments made with white rats; with ants that are undisturbed, or at least upset as little as possible, we have a chance of catching their true reactions to the maze.

Near my house in the country, I own one of the woods typical of the Ile de France, full of ants and birds. It is my open-air laboratory, where I can set up, free from disturbance, all sorts of apparatus and carry out all kinds of experiments. I once sat down at the side of a sloping track the ants had followed faithfully for seven years and pondered as to how one might demonstrate collective learning among them. It had to be arranged that the same ants should go along the same track for several hours together, and even, perhaps, for several days. Now, here we were helped by the subject of the trials; because some old experiments made by Wellenstein and Otto show us that red ants are particularly "faithful to place" (*Ortstreue*), that is to say they follow the same track for months and even several years together.

I then set about canalizing the workers' movements by means of that magic substance industry has so fortunately put into the hands of the myremecologist: fuel oil. Red ants hate it and you can make them go practically where you like with the help of cotton wool or cardboard soaked in this substance. The

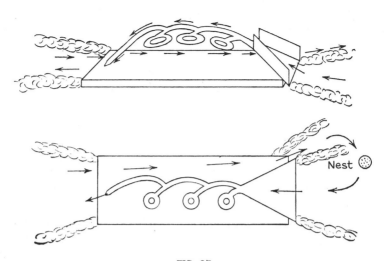

FIG. 37

Loop maze used with *Formica polyctena* (after Chauvin). *Above,* lateral section. *Below,* plan. Here and there strands of cotton wool soaked in fuel oil.

repellent substance is not the fuel oil as a whole but a non-volatile ingredient that it contains; and even in the open in full sunlight, when neither the cotton wool nor the cardboard smell of oil (at least to human nostrils), they are still repellent to ants.

I forced them, then, to go in dense crowds to a spot where I had put a maze and, the height of luxury, I separated them into workers going up and workers coming down, by means of barriers better explained by a figure than by long explanations (see Fig. 37). I used quite simple mazes (Fig. 38) and the one that gave the best results was the loop maze (Fig. 38, I). These loops are necessary, or at least very useful, because when ants crowd into an over-narrow cul-de-sac they pile up in such a way that many of them drop to the bottom of the maze and there is no longer any question of following them individually. With the loop they carry on in relatively good order.

I then devised a simple test: I counted the ants entering each of the three loops and established the percentage which, on getting back to the main long track, *continued in the right direction;* that is towards the exit, instead of going back on their tracks as the cul-de-sac treacherously invites them to do. Another advan-

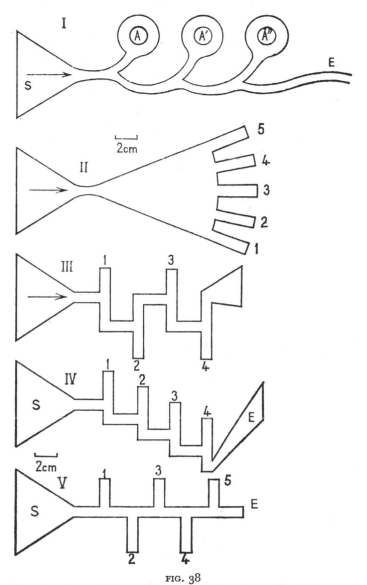

FIG. 38

Mazes used to study learning ability in *F. polyctena*. No. I gives very good results. The ants cannot learn No. III. S. Start; E. Exit (after Chauvin).

tage of the loop maze is that the main problem is presented three times in succession.

By the liberal use of fuel oil I steered my ants towards the maze. I had difficulty in choosing some twenty or thirty of them at random and in observing everything they did. Particularly as the stream of brown workers on the white floor of the maze has something hypnotic about it. I was fascinated by them: no one had ever done such an experiment before; and so, particularly with ants, anything could happen. What, in fact, would it show?

Well, that they learnt something was incontestable. Almost immediately the preliminary agitation died down and the workers split into two lines: the first, who took the direct path, was composed of those who by chance found themselves on the right-hand side of the passage (that is if the blind alleys were on the left). The second stream went into the first cul-de-sac, "looped the loop" and, coming out, again split into two portions, one going towards the exit and the other towards the entrance. As time went on the numbers going the right way (towards the exit) increased. Often, moreover, ants which had learnt how to deal with the first loop did not even try to enter the following ones. Four manœuvres worried them to some extent; these were: moving the loops to the right when they had been on the left, reversing the direction in the maze, using a new maze instead of the old one, and finally moving the maze to a completely new environment some ten metres away. The percentage score was still very high, well above that obtained at the beginning of the trial. This allows us to draw several surprising conclusions:

There was no scent trail left in the maze, such as is put down by many ants (e.g. the wood ant *Dendrolasius*); otherwise, offering them a new maze would have disturbed them very much.

The sensory reference points, whatever they may be, occur in the maze environment itself rather than in the outside world; otherwise moving the nest ten metres away would upset the ants.

A lesson learnt can to some extent be transferred, for example from left to right or from forward to backward directions, contrary to what Schneirla maintained.

All this might make one feel optimistic, but looked at more closely there are some considerations that bring us back to earth.

"*Photohorotaxy*"

First of all, the learning we noted was only on a statistical basis. The ants never learn, even after a week, to avoid the loops altogether; even after this lapse of time nearly all the ants go through them several times. It then struck me that there was a great contrast between the white floor of the maze and the brown soil of the forest; a reaction that Kalmus discovered at one time, saddling it with the strange name of "photohorotaxy". . . . If you paint black bands on white paper all insects with eyes tend to follow the line of the border between the two contrasting stripes. And might not our ants in the maze also be subject to this automatic reaction? It would explain why they found it impossible to correct mistakes, as well as the attraction the loops had for the file of worker ants on that side. . . . A way of finding out was to do away with the contrast. Flour paste is not repellent to ants, so using this I stuck earth and twiglets to the floor of the maze. This made it look awful, but the result was startling: direct runs through the maze, with the ants leaving out the loops, increased from 20 per cent to 80 per cent.

Progress backwards

The above is a good example of the steps backward which I mentioned just now. . . . Good Lord! I should have thought of photohorotaxy before. There is no doubt that all learning activities are upset by it. But is that an irremediable drawback? I do not think so; firstly, because even in the maze there is a gradual improvement, which it is difficult to attribute to any cause other than learning: only it is less marked than it should be. Other facts cannot be explained by photohorotaxy: for instance, the step maze (Fig. 38) is easy for ants and by contrast a maze of alternating wrong and correct ways is of insurmountable difficulty (on the other hand it should be noted that cockroaches learn such a maze very easily).

The result of one experiment much intrigued me and I am tempted to believe it showed almost a fundamental understanding of the problem in our subjects: this was the short-cut experiment (Fig. 39). *All* the ants passed through the short-cuts the right way, that is from A to B and *never* in the opposite direction.

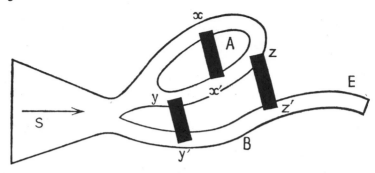

FIG. 39

Experiment with *F. polyctena* in "short-cut" maze. The ants starting at S enter the loop A in considerable numbers, where three short-cuts (in black above, though in practice they are the same colour as the maze) allow them to regain the main track. All the workers move from x to x', from y to y' and from z to z' and none in the opposite direction E, exit (after Chauvin).

One could have sworn they knew which was the correct direction, though the way the problem was posed and their own internal machinery hindered them from following it. If photo-horotaxy alone was concerned one cannot see why the passages should not have been used in both directions.

The role of kinesthesia

As to the senses used by ants to find their way around, it is apparent that sight is involved in photohorotaxy. But another sense (a group of senses) is also concerned: one called kinesthesia, or the remembering of movements and positions of the body. For instance, if you suddenly double the size of the maze, the ordinary observer would not think that any essential factor had been changed. But in this last case ants that knew how to thread the smaller maze were at a loss in the bigger one and made as many mistakes as at the start. This is the classic test, made with rats, that leads to the hypothesis of a kinesthetic sense.

The "kind of equipment"

I made another experiment which shows the curious differences between workers, according to the tracks under observation. It was to move the maze from one of the ants' tracks to another: it has to be relearnt from the beginning; there is no exchange of information between workers using different tracks. But a strange phenomenon was seen: there was a *considerable difference in learning ability according to the track used*. It seemed as if the ants on certain tracks were considerably "sharper" than on others and learnt quicker and better. Moreover, memory differs according to the track used. On some the performance of workers does not fall off after four days (which is a long time for an insect). Whilst for others the memory has already deteriorated after two days. . . .

Now you see how a few bits of cardboard and a tin of fuel oil help us to penetrate more deeply into the minds of the red ants. But I thought up yet more ways of teasing them, some of which at first filled me with enthusiasm, but finally proved disappointing. Others showed a welcome capacity for solving certain special problems.

Climbing a precipice

I placed two wooden blocks on a track touching one another and then I gradually moved them apart. In order to get over the blocks the workers stretched out as much as they could from one to the other, but eventually the gap proved too wide: they had to fall into the gap and then climb up the next block in order to continue on their way. On one of the blocks I then put a few twiglets; all they had to do was to tip them down the

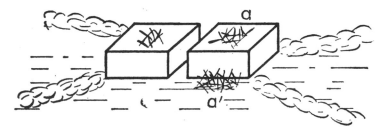

FIG. 40
A simple problem the ants cannot solve. Two blocks of wood are placed on a track so as to leave a small space between them. A little pile (a) of twiglets is put on each block. All the ants need to do is fill the gap with the twiglets in order to cross easily from one block to the other. However, they throw the twiglets off at the side (a') (after Chauvin).

gulf, fill the gap, and cross over. Alas, nothing doing! They took the twigs and threw them to one side but not down the precipice. I had to resign myself to the situation: although they were handling twigs all day long, to cross an abyss on a pile of twigs *over a path* is not a method ants can accept.

Making a scaffolding to reach prey

Everyone has heard about the ape that piles up boxes in order to get a banana placed out of its reach. I thought up an experiment to see if ants were as intelligent as apes. Over the top of an ant-heap I suspended a succulent dead wasp from a thread, adjusting the height so that it was not very far above the ants. And I saw a little pile of twigs formed beneath it. . . . But looking closer I had to admit after all that ants were not apes. When the distance between the food and the surface of the mound only slightly exceeds the length of a worker what happens is that the ant can grip the food with its front legs, but at the same moment its posterior legs are gripping a twiglet and that raises it up a little. If this action is repeated several times by other workers, a little mound is piled up. But it is not intentional.

Lubbock maintained that when *Lasius* were shown pupae placed on a platform a little above their reach the ants built a little mound which enabled them to get at their young; Gœtsch witnessed the same thing. I do not doubt the observations of these two brilliant myrmecologists; but I do ask if their interpretation of the phenomenon is correct. That is to say, whether the ants' solution to the problem is of the same order as that of the chimpanzee building up boxes to climb on and pick the bananas hung out of his normal reach. But I doubt it. . . . Do not forget that we are dealing with ants; it is just when their actions seem most like those of the higher animals that we must be cautious: the underlying mechanism may well be quite different. . . .

Lining a tunnel

My ants managed to pass through a Fontainebleau sand ridge by means of a lined tunnel. This white sand, almost pure silica, runs like water when it is dry, and must present almost insurmountable problems to burrowers. At least unless they line the galleries, as do human miners. I investigated just such a tunnel and found that it was well and truly lined with twiglets laid lengthwise along the tunnel. This is what happens. When the substratum gives way, or when it is wet or sticky, the ants bring up twiglets and lay them on it. But when the twigs are in place they dig beneath and gradually bury them. At a certain depth the sand is a little damper and there are then enough spaces between the twigs to allow the ants to pass through quite easily. From which I conclude that the ants line their galleries, *but not in the way our miners do.*

FIG. 41

A common and automatic manœuvre among ants; they bring twiglets and general detritus to cover a sticky surface, which can then be crossed. Detail of this practice in the Argentine ant (after Ghidiri-Pavan, 1957).

Climbing a wall

I surrounded the nest with a very high, continuous wall of cardboard or aluminium. I was surprised to get three different reactions from three different nests. One colony simply deserted the nest, crossing the wall simply to establish itself fifty metres away. It was an unusual reaction, because ants can easily scale high walls, though they do not like doing so. Another colony bored a hole thirty centimetres above soil level (no doubt in a weak part of the cardboard) and promptly re-covered the mound, which I had destroyed, with twiglets. The third nest was surrounded with aluminium: here the ants dug a long subterranean tunnel through which they brought biggish twiglets to refurnish the mound (which, again, I had destroyed). It was rebuilt in a few days. This plasticity of reaction is all the more curious in that the solutions found are not habitual. It is very rare in nature to find twiglets brought in by underground passages. They usually arrive at the nest surface by the open-air route.

CONCLUSION

What is an ants' nest? A comparison with man and his works

THE PICTURE I have just given of the ant world may appear a little confusing to some of my readers . . . I am well aware that my book, in its lack of continuity, has some resemblance to those "Marvels of Nature" treatises that were so popular in M. Buffon's time and even a little later. But there was nothing else I could do: I could not present the subject as a unified whole when there are still so many gaps in our knowledge.

Nevertheless, is a synthesis impossible, however provisional it may turn out to be? Not altogether impossible, provided it is presented as greatly hypothetical. Some people will un-doubtedly find I am skirting too close to the borders of science fiction: but results of advanced research often do resemble it. And to the devil with those prematurely old persons whose self-styled caution stifles the imagination (or would stifle it if they had any!). No, there is nothing wrong with the term "science fiction", provided we consider the theories put up as provisional tools, as quickly to be taken up as dropped, according to circum-stances. I have read somewhere, I no longer remember where, the story of the director of a research station who made his staff frequently read such works of scientific prophecy; as no doubt you have guessed, he was American. It may have been naïve, but he thought it to be the best way of loosening up his staff's thinking.

The "most intelligent" ants

All through this book I have emphasized the incurable stupidity of isolated ants. In all laboratory behavioural studies of orientation, or of sense discrimination, they show the same pitiful results; they do a thousand times less well than a monkey or the most obtuse mammal; they behave like very rudimentary and "unintelligent" machines. This is hardly unexpected in view of the smallness of their brains, which contain only a few nerve endings, even though these are smaller in insects than in the higher animals. An idea immediately strikes one: why not try working with the biggest ants? they should have more sense than the little ones! But this is not the case and the big *Campo-notus* proves neither more nor less stupid in the laboratory than the little red ant. What is more, a myrmecologist would be very surprised if you told him that the biggest ants must be the most intelligent. To say nothing of size differences within the colony, which can be enormous. We have seen that a huge soldier ant may have a very small brain, inferior even to that of an average-sized worker. In comparing species it is equally false to suppose that any relationship exists between complexity of behaviour and size. The huge African *Paltothyreus* are not remarkable for their intelligent behaviour, any more than the gigantic *Dino-ponera* or those Titans of the ant world, the *Myrmecia*, who seem to be as stupid as they are ferocious.

Unquestionably the most striking behaviour is found among *Atta*, who cultivate a fungus; the red cattle-raising ants; the slave-making ants which enslave several different species: or man's ingenious enemies, *Solenopsis* and *Iridomyrmex*. Now all these ants are medium-sized or small: *Iridomyrmex* is minute. They have, however, a striking characteristic in common: the *enormous numbers* of individuals in each nest (the champions are the driver ants, *Anomma*, if it is true that, as Raignier says, the population of a colony can be more than twenty million). In addition, not only do the nests of these species have high populations but also they all have a tendency to form a colony of nests. We shall see (on p. 206) the importance of this factor to behaviour.

Thus the most intelligent ants, or rather ants' nests, are the most numerous. We can now see the steps Nature has taken to escape from the rigid mould of the insects' small size. We learn from neurophysiology that an intelligence cannot be developed beyond a certain point unless there are a minimum number of nerve endings. It is impossible to increase the number of these endings if the size is not increased as well, because the nerve centres cannot exist below a certain size. It would be different *if the small brains could be added one to the other* or multiplied in some way. And such a thing is only possible in an insect society. In such a case the mind would take a big step forward, for the combined brain mass of a million foraging ants could easily reach 100 grams; that is to say a perfectly reasonable number of nerve connections, considerably higher than that found, for instance, in the brain of a rat.

But why do they make so many mistakes and do such strange things?

Nobody denies that ants' achievements far outdo those of any other animal: for instance, no monkey is capable of cultivating a fungus, of keeping cattle, of storing grain, of making slaves, nor of getting disgustingly drunk, as red ants do with the intoxicating secretion of the *Lomechusa*. Man alone can do that. For a long time, perhaps from the time of Solomon, observers have noticed it, and been disturbed by it. Some of them have not hesitated, even in modern times, to credit the "spirit of the nest" with the qualities of foresight and reflection, in short— we might as well give in and say it—"intelligence", comparable to those found in human behaviour.

Nevertheless a doubt persisted, and as the number and accuracy of observations has increased, the earlier position has been driven into a corner. We know the leader of the opposition quite well: he was called Etienne Rabaud. White-haired, malicious, with the smile of an old Lucifer beneath his thin white moustache, nothing pleased him more than to clamp the work of a colleague into the pillory of ridicule; and the young research workers, of whom I was one in those far-off times, used to shriek with laughter as they quoted the biting sallies of this ingenious old man. All the same, he was a wonderful observer, full of skill and understanding. And he quickly saw the incredible foolishness that appears to characterize ant behaviour: for instance, when it is a question of digging a hole, a worker will put a pellet of earth at the side of the cavity it has just made; and go away. It may well happen that another ant will come along, pick up the pellet and pull it back into the hole. The same thing happens when red ants are carrying twiglets to their mound. I have watched for several days a hundred or so of them individually as they carried construction material, to find out what they were doing: at times I have been maddened by their behaviour, the utter uselessness of their comings and goings, their hesitations and wasted time: once or twice, in the depths of the forest, I have caught myself shouting with exasperation: "Put your twig down, you fool! It will do as well there as anywhere else!"

From all this Rabaud drew a conclusion that has remained famous for its very outrageousness. According to him social insects are only social in appearance and only because a vague inter-attraction draws them together. In reality they live *as if they were solitary*, without in the least attempting to co-operate with one another. Thus an insect society is merely a *"conglomeration of solitaries"*; ants, for instance, are no more social than are iron filings drawn together by a magnet.

Order arising from disorder

One can easily imagine the violent arguments that raged round such a paradoxical statement. In the 1930s neither those who maintained this idea nor those who denied it distinguished themselves by an excess of logic. And how much sour comment has been poured over all this by our learned universities, as happens so often in our world of science. Lucretius sang of "the calm temples, loved by the learned"!; the temples alas, are only calm when the learned are not there!

The heat of controversy hid one blindingly simple fact from Rabaud: that the mound *is finally made, and very quickly too*, and can reach a considerable size. So order and accomplishment have arisen from absurdly foolish individual efforts. Can that be true?

We have only come to understand it quite recently. Since the emergence of cybernetics and the work of von Neumann, to be precise. Many people know this famous German mathematician at least by name: he was one of the first to show politicians that higher mathematics could be of some use; for instance, in organizing an improvement in the war effort, positioning a transport fleet so that it continuously offers the minimum target area to enemy submarines, etc., . . . but also, and above all, for producing more and more powerful calculating machines and imitating living creatures by means of cybernetics.

I cannot discuss this new science here, but almost everyone has heard of it. It attempts to study the numerous "living prototypes" (as the Americans say) that are found in nature: for instance, the frog's eye following a living prey and estimating the distance at which it can best be snapped up; the sonic device of the bat, so delicate that it leaves our own radar detectors far behind. But, above all, what von Neumann was seeking to understand was the working of the most prodigious piece of living apparatus known, one we admire without yet understanding: the brain, the human brain. And that is what, paradoxical as it may appear, brings us back to ants.

To be exact, von Neumann and the other cybernetics experts (Wiener, McCulloch, and Ashby) were not trying to make a brain—at least not at this stage—but to build automata capable, for example, of appreciating the position and distance of a target, of getting the apparatus to home on it, and of overcoming obstacles *en route*; capable, in short, of behaviour in what might provisionally be called an "intelligent manner". In this field fantastic results were soon obtained. But what interests us here is not so much the results obtained as some of the rather strange observations made by the workers on the way.

The breakdown problem

Making one of these automata needs a large number of electronic components, condensers, transistors and relays of all possible kinds: and thus there is more than a chance, alas, that some components will fail. How is it, then, that in the human brain, the elements of which are thousands or hundreds of thousands of times more numerous, failures are so rare and can mostly be corrected automatically? The brain elements, or neurons, moreover, are at least as complex in their working as are the electronic components of an automaton. For some little time now, neurophysiologists have come to realize how nerve cells and fibres work, by inserting minute electrodes into them; ultra-sensitive apparatus then measures the current arising from the beginning or flow of a nerve impulse. By means of these instruments one might expect to find that a nerve impulse starting in a peripheral sense organ of the body always follows a more or less constant course in the cortex, to end up in the same, given group of cells. This is not at all what happens.

"Disorder" in the brain

All that the neurophysiologist finds is an extremely irregular flow of electric impulses, following one another in no particular order, diminishing in one place and increasing for a while else-where. The flow can follow almost any route without causing disturbance. This is why one can excise considerable parts of the brain cortex without stopping its action. This is what so sur-prised the old authors: can the message no longer follow this route? Very well, it can go some other way just as well. This is something quite different from our machines, or, more exactly, from the machines we used to make. Now, in the ants' nest, what happens if you destroy half the workers? Nothing, or almost nothing; the implacable ant-machine soon draws on several thousand more units from its depths, who carry out the task as well as before—or as badly. Whence comes this flexibility and this absence of mechanical failure which seem to be connected, in the ants' nest as much as in the brain, with a certain disorder?

What kind of relationship, then, exists between these ele-ments? If each one was connected up with its neighbour by an unalterable interrelationship, like two cog-wheels, as it were, we would no doubt achieve order and regularity, but not adaptability or absence of breakdowns. But there is no provision for adaptability nor for mechanical failure. It is worth remem-bering that one grain of sand can bring an entire piece of machinery to a halt. Not much progress can be made along these lines. Vaucanson's automata would be the limit of development in this field. However, electronic engineers have reached a stage of development where they have succeeded in creating machines whose "internal logic", as it is called, greatly simplifies the construction of "supermachines"—which, indeed, might not be possible without it.

Majority organ, Sheffer stroke organ

I will try to explain these terms by borrowing largely from the works of one of my pupils, Jean Meyer, both physicist and zoopsychologist; which is to say he is one of those fortunate beings who understand something of mathematics and is not terrified, as I am, when faced by a yard-long equation: but do not worry, I will not inflict them on you either. Two of the most important components of modern automata are the *Sheffer stroke organ* and the *majority organ*. At this stage no one needs to know or understand exactly what is inside the "black box", which is the apparatus; it is always a combination of wires, transistors and simple bits of equipment by means of which it is easy to get from the "organ" the simple reactions required of it. It is sufficient to state that the *Sheffer stroke organ* allows a current to pass unless two inhibiting impulses both reach it at the same moment by its two entry lines. As to the *majority organ*, this has three entry wires and one exit one: it allows nothing to pass through it unless two out of the three entry lines are activated (the majority).

Here then are two simple pieces of apparatus which could play the part of the simple elements in an ant's brain. For instance, let us suppose that a worker is carrying an ant corpse to the "cemetery", that is to say the place where they assemble the colony's dead ants and rubbish. The three entry lines of the majority organ could, for instance, be three sensory stimuli aroused by the corpse, such as smell (experiments have shown that the smell is the strongest stimulant determining its transport by the workers), general appearance and rigidity. The transport is undertaken if two out of three of these stimuli are present. Then the current passes through the majority organ and can go through to the Sheffer stroke, which will oppose its passage only if two contrary stimuli reach it at the same time (for instance, strong stimuli inviting the ant to do something else such as seize a victim or a twiglet). I advisedly say two and not one: you will see in this way how the bias is incorporated into the system, thus giving it its flexibility. In the last case the ant abandons the body. But it may only move a short distance from it and return when the contrary stimuli have died down (how often have I watched such returns); then the corpse stimuli impose themselves again and the ant starts to carry the body once more.

"What nonsense!" you might well say. "All that is very naïve and could be put much more simply without bringing in all this boiled-down cybernetics." Not at all, but here we must put our faith in Meyer's calculations. I think we can go one step further without being suffocated by hydra-headed mathematics. Let us suppose that three stimuli reaching the majority organ are all positive, all the same, and of necessity linked together (for instance, in the case of a living victim, its appearance, movement and smell cannot be separated one from the other). Then the probability of an error or faulty working of the system in two lines together (the majority) is less than one-sixth: this means that the probability of error or faulty working (even if one of the entry lines is functioning badly) is less at the exit than at the entrance: *the majority overcomes one member voting against it.* The majority organ is capable of distinguishing truth from falsehood. Obviously in the case where the three stimuli are independent (not connected with each other) the chances of incorrect functioning are, on the other hand, greater at the exit than at the entry.

Now, naturally, the three lines of the majority organ can perfectly well be represented by three ants stimulating a fourth with strokes of the antennae, exchange of food, etc. And the three stimuli need not be effected at the same time because ants are provided with memories. In other words, ants do not react to a "background noise" where all the stimuli are independent, but can distinguish the significant stimuli of a complex situation even when they are weak. The separation mechanism can be made yet more powerful, because the first ant notified by its three companions can join with two others, similarly informed, to inform a seventh ant; and thus the probability of error at the exit of the second sub-group, made up with these new recruits, should be correspondingly smaller than at the first group's exit. Von Neumann has written somewhere that the *probability of a co-ordinated working* of such a system increases in ratio with the *number of elements* (neurons or ants) it contains. There is thus a considerable chance that a large group of ants will form and proceed to carry out the same actions. One can easily imagine the end of such a chain of action becoming progressively better informed and in turn instructing the start of the chain; and that the current running through the closed circuit would become still quicker and more efficient. We see many examples of these closed chains, as during trophallaxis, when the ant who has just given food solicits it again, almost immediately, from the one to whom she gave it!

Division of labour or co-ordination between sub-systems

The picture I have just drawn of the basic mechanism of an insect society is obviously very rudimentary. We all know there are specialized sub-groups, or sub-systems as the cyberneticians would say. For instance, the nursing group in an ants' nest, or more generally the "internal workers", as distinct from the "external workers". Now each of these sub-systems visibly adapts itself to any given situation: for instance if an ant roadway is barred, the ants are perfectly capable of making a detour, even if only after some trial or error. But what do we mean here by the word "adaptation"? It should mean a series of successive changes of tactics which only stop, as Ashby says, when the system reaches an "ultrastable" state within the limits set by the physiology of the individuals in question. Now, contrary to what Rabaud thought, it is very evident that these systems are interconnected in various ways. For instance, if the nurse ants stopped feeding the queen and brood, all the colony, including the foragers, would soon be in jeopardy, owing to the progressive aging of the workers. As a matter of fact, it would be very difficult for the nurse ants to cease working, because the foragers bring in a continual supply of food which passes from crop to crop and gives rise to an "internal food pressure" which tends to get the food distributed to others.

But, and this is the cause of Rabaud's illusion, *if these sub-systems are interconnected, the connections are not very firm.* Their connections are weak—at times very weak; examples of obvious unco-ordination spring to light all the time. Ashby's calculations make it possible to understand why: it is inevitable, since weak connections, with their train of faults, are the condition *sine qua non* for the working of systems made up of the ultrastable type of sub-systems. I cannot give an account of Ashby's admirable paper here; but a piece of fairly simple reasoning will do instead. Each sub-system has a very different equilibrium: the foragers lead a very different life to that of the nurse ants. Thus, as Meyer says, "it is clear that the less any individual adaptations react on one another, the more quickly will this general adaptation be achieved, thus avoiding a sub-system's readaptation to a given situation causing an essential variable in a neighbouring sub-system to exceed its physiological limits, thus necessitating a further readaptation of this sub-system." Let us consider a population of balanced elements that from time to time get out of equilibrium; if it is desirable that the equilibrium be re-established as quickly as possible the best procedure is to let the elements oscillate separately. If these elements are connected to each other, however slightly, there is a considerable probability that the equilibrium of the whole will be delayed, particularly if the elements out of balance are not identical. You cannot regulate one part of a watch without instantly affecting the rest of it; in the case of an ants' nest one can, and even should, regulate each part independently of the others.

Is the ants' nest a giant brain?

One must remember that von Neumann, McCulloch and Ashby had little interest in ants. We have seen that their study concerned automata and the brain, and from these they made the calculations from which cybernetics was born. This, then, is the moment to put the question, "Is an ant community only a loosely connected giant brain?" No, we will not go as far as that: but we cannot help noticing a strange similarity. We now know that there is no "spirit of the hive", or "spirit of the nest", but a hypercomplex of sensory interactions weaving a tight if sometimes random network of interactions between the ants. It is this network as a whole that constitutes the ant colony. Is that not how the brain works too? Who was it that compared it to a huge central telephone exchange without any telephone operators? And where does intelligence or consciousness reside in all this? We do not know; we can only suspect that the question is badly put, as it is when we speak of the "spirit" that animates the ant colony.

We must also add that electronic engineers have now constructed circuits on which the different elements are joined to each other by as many connections as possible. If the connections are both numerous and random the *whole network* then has certain properties which remind one of the brain (or of ants); for instance, the stimulation of an outer part of the system directs an impulse to another part along paths which are not necessarily always the same. And if you remove a piece of the network, at random, nothing very much happens: the impulse continues to flow simply by using other paths. For instance, you can train a rat to press a lever when it sees a light go on, thus getting a reward. Naturally such training implies the use of the visual centres of the brain. Very well, let us use a large scalpel and remove nine-tenths of this area, from anywhere you like so long as you leave a tenth part; the rat will not lose its sight or forget its training. . . .

Once more like an ant colony. . . . In the course of studying the technique of ant-raising that Gösswald perfected it was found that you could dig out at least twenty litres of nest material, brood and workers from a nest. And what happened? Strictly speaking, nothing: if the nest was sufficiently big it continued to live as usual.

"What hypocrisy!" some readers may say. "First of all you deny that an ants' nest is a brain, and then you take the opportunity of quoting every instance that suggests it is!" Yes and no: this attitude does no more than reflect my own uncertainty. I well know that an ants' nest is not a brain, but that does not mean that the basic principles of their organization are not similar, or that the study of the ants' nest cannot teach us something about the brain, and *vice versa*. That would be a most unexpected result of the study of myrmecology. . . .

Are ants still evolving?

Since we are discussing such controversial matters I might as well be hanged for a sheep as a lamb! I think I can say that for fifty million years at least there has been no sign of any morphological change in ants. And everyone tends to repeat what he has read in all the books on the subject of instinct in Hymenoptera: that instinct which, in spite of its marvels, is so absolutely rigid, so strictly stereotyped, that one cannot see how it could change. . . .

But let us beware of opposing ideas too universally accepted. I do not know if ants are evolving beneath our eyes, and I don't really see, up to the present, how I can find out. But as to the absolute rigidity of ant behaviour and the impossibility of any change, I have kept a sharp eye on ant biology and have seen how clearly certain facts emerge, facts patent to everyone and yet, in my opinion, given far too little importance.

I believe that ant behaviour can include disturbing cases of imitation. For instance, there is the case of *Lasius umbratus* of which I have already spoken; this underground ant hates the light, yet does not hesitate to follow *L. fuliginosus* to the tree tops in full daylight when invited, and, moreover, then brings up its host's brood. Let us also remember the very strange phenomenon of the multiple slavery of *Formica wheeleri*, whose slaves *neorufibarbis* help in their slave raids, although in their natural habitat *neorufibarbis* never make slaves, and whose other slaves of other species raise the brood and attend to the nest during this time. Their behaviour has not been much studied, but it may be that they have learnt some of their masters' habits.

As regards instant adaptation to unique situations never before encountered, man's fierce enemy, *Solenopsis saevissima*, the fire ant, springs to mind. There is something astonishing in the way it escapes from insecticides and poison bait. It is the same with *Iridomyrmex humilis*, the Argentine ant, which often mocks at our methods of control. Pavan records that one day he placed a ring of insecticide on the soil around a tree, thinking he would discourage the ants from pasturing their aphids there. Alas, a few days later the greenfly were as numerous as ever; the ants had merely made a tunnel under the poison ring . . . a difficult matter to explain.

But there is one thing rather more important than these somewhat casual observations—for, in my view, they have never been closely enough examined—and that is a phenomenon which has been observed only four or five times to my knowledge, and incompletely even then: the matter of federations of nests.

Super-colonies: what do they indicate?

Let me repeat a remark I made at the beginning of this book; the most "intelligent" ants are not the biggest but those that have the most highly populated nests. What happens if several nests associate together to form a super-colony of some tens of millions of individuals?

The reply is both simple and disappointing. Such super-colonies do exist, but they seem to be rare; what is more, they have not been studied in depth; authors have not realized the importance of the phenomenon. For my part I only know of four detailed studies: Forel (1911) on *Formica sanguinea*; a much older one by MacCook (1877), very long but very vague; another by Wasmann (1909) and especially two more accurate ones: by Stammer (1937) and by Raignier (1952), dealing with red ants apparently belonging to the species *polyctena*. But first let us get rid of an objection: myrmecologists have known for a long time about "polycalism", that is the ability to form daughter colonies linked to the parent colony, and that it is not at all rare: it occurs, for instance, in the Argentine ant and the fire ant: it is indeed one of the reasons why it is so difficult to control these ants. Then why be so interested in super-colonies? Because of the *population factor*, which plays an astonishingly important part in the cases I am about to expound. According to the few data we have, it is only in the genus *Formica* that really enormous numbers of individuals are found (to tell the truth I expect that similar phenomena exist in the super-colonies of *Solenopsis* and *Atta*, but that aspect has not been studied at all).

I greatly regret that MacCook's work was so imprecise, because his super-colony was the champion of all such noted so far. It was found in the outskirts of Philadelphia and contained from 1,300 to 1,800 nests. If one puts the nest population per colony of this species (*F. exsectoïdes*) at 300,000, which is low, one gets a total of some 450 millions of ants found together on a few hectares. The super-colony observed by Wasmann was *rufa*, on a more modest scale and situated on the side of a hill near Derenbach in Luxembourg; in it some fifty nests were spread over an area 200 metres long by 70 metres wide.

Stammer's super-colony was more imposing: he found it in the Gulitz forest in Mecklenburg; it had 58 principal nests and 31 secondary ones (about 2,900,000 ants), joined by paths, of which he made a map. The distances totalled some seven-and-a-half kilometres.

Raignier's super-colony was on a plateau between Namur and Dinant and covered 30 hectares of ground, encircling a hill with a network of paths and mounds; it contained about a hundred nests (50 million ants).

Finally, the big super-colony observed by Forel belonged to a different species, *F. sanguinea*: it had some forty nests all in communication with each other, along a line measuring more than 150 metres.

I do not know of any other recent observations on these ant federations, but I myself am now about to start a more detailed enquiry.

Characteristics common to red ant super-colonies

The various observations made by authors, though too meagre for my taste, do enable one to find a few points in common:

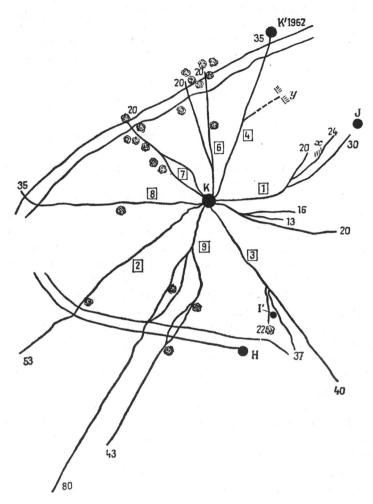

FIG. 42

The *F. polyctena* nest studied by the author for fifteen years: it gave rise
to twelve daughter colonies. The numbers within small squares are
the track numbers; the numbers at the ends of tracks are the lengths
in metres. K, main colony; K', daughter colony; J and H, other nests
not descended from the main nest; X and Y, temporary nesting
places. The irregular black circles are pine-trees.

1. One absolutely essential point, noted by all observers: The ants show no hostility to workers from other nests in the federation, even if these are taken from nests at several hundred metres' distance from one another. And Heaven knows, this seems extraordinary to a myrmecologist!

2. In these colonies, where the queens are perpetually renewed, the nests stay in the same place for decades at a time: they can be plotted on a map together with their main paths, persisting from one year to the next; this is particularly easy as they are flattened and free from herbage (Stammer).

3. Man cannot penetrate the federation's territory, unless he takes special precautions. It is impossible to stand still and difficult to move about. Stammer mentions the numerous ants which fell from trees into his clothes and Raignier says much the same thing. The aggressive spirit of the super-colony seems very marked.

4. The grouping is not completely homogeneous; there are groups of nests joined by wide, well-marked paths and other groups where the tracks between the nests are much fainter.

5. The different nests are not all of the same size, and there is always one "giant nest". Raignier was much impressed by an enormous ant mound, the "star-shaped nest", more than a metre high with a basal diameter of three or four metres and a circumference of fifteen metres from which fifteen great paths diverged. It is curious that the giant nest described by Stammer at the middle of his super-colony was about the same size: a metre high and nine metres in circumference. This is about all that is known of these immense federations; except, we must add, *the noise* that so impresses all observers: it is the sound made by millions of tiny claws moving over the leaves, reminding one of a heavy shower of rain. Questions accumulate to which there is not as yet even the beginning of an answer. For instance, how are such multitudes of voracious ants fed? Then are there exchanges between nests; and, if so, of what kind? We already know that such exchanges take place. Kneitz has studied them in mother and daughter colonies of *F. polyctena*; I myself have fed radio-isotopes to different colonies of *polyctena*; this made it easy to note how one colony after another, even when fifty metres apart, become equally radioactive, even colonies of *F. rufa* as well! But who will tell us the meaning of these

exchanges, who will chart their probable reciprocity for us? Remember that in ants we have already found the equivalent of agriculture and stock-raising: in super-colonies why should we not find the *equivalent of commerce*? It is not impossible, if we define this term as the *reciprocal exchange of goods*. Do not forget that Kneitz has remarked that everything is exchanged: brood, queens, food and building material.

In addition, the highly aggressive nature of these ants shows them to be the most susceptible to stimulation. How, then, would a federation react to experimental interference, such as the destruction of a nest or its shutting off from the main federation by means of a repellent barrier? For it is when the network of sensory intercommunications is crackling at every intersection that new solutions (though of an ant and not a human kind) will emerge under the eyes of the observer.

I ask myself in all seriousness whether ants, already so far advanced along the path of complex behaviour, are progressing still further in the recesses of their federations. On this note of doubt tempered by hope I will end my book.

Les Sources
August 1968

INDEX

Acromyrmex, 82, 83–84, 86
Acropyga, 100
Alarm substance, 153–155
Amorphocephalus, 130–132
Anal glands, 160–161, 162
Antelopes, 130
Anergates atratulus, 20
Anoplolepis, 97, 157
Anomma. See *Dorylinae*
Antennae, 144–148, 175
Ant-heaps. *See* Ant nests
Ant nests, 29–30, 34–40, 43–52, 87,
    104–111, 149, 196, 204–205,
    206–210
  compound nests, 135
  construction of, 105–111
  fungus and, 78–87
  humidity and, 50, 51–52, 81
  shape and supervision of, 35–37,
    38–40
Ant plants, 107
Ants:
  brain and senses, 163–195
  brood, 111–113
  chemistry, 158–162
  classification, 10–15, 28–29
  communication, 143–149, 150–153
  cultivators and harvesters, 79–88
  destroyers, 69–79, 89–91
  differences in individual reaction,
    122
  division of labour, 120–127
  herds, 92–103
  nest as possible brain mass, 195–
    210

  nests, 104–111
  numbers in colony: their signifi-
    cance, 124, 196, 202
  origins, 22–25
  parasite guests, 128–135, 140–142
  physiology, 15–20
  queen-making, 113–120
  red ants (utility, nest, heating,
    food), 26–69
  slavery, 135–139
  territory, 155–158
Ants (references to habitat):
  Africa, 13, 14, 73–74, 79
  Algeria, 14
  America, 71, 76, 89, 98
  Argentina, 97
  Australia, 13
  California, 98
  Ceylon, 23
  Chile, 88
  China, 102
  France, 107
  Indonesia, 15
  Italy, 105–106
  Java, 100, 101
  Kenya, 98
  Siberia, 30–31
  South America, 107
  Texas, 89
Aphaenogaster, 166
Aphids, 15, 20, 59, 61, 69, 92–102,
    138, 144, 155, 156, 168, 178,
    206
Aphis sambuci, 94
Forda formicaria, 92, 99

Aphids—*cont.*
  as pests, 97–98, 102–103
  predators of, 96–97
Apocephalus, 84
Argentine ant. See *Iridiomyrmex humilis*
Ashby, 199, 203, 204
Atelura (silver-fish), 128, 131
Atemeles beetle, 131, 132–134, 142
Atta, 14, 18–19, 20, 78–87, 101, 124, 128, 156, 177, 196, 207
  Atta sexdens, 78
  Atta cephalotes, 80
  Paedalgus termitolestes, 14
  fungus and, 79–87
Auto-amputation, in ants, 22
Autuori, 79, 80–81, 85
Ayre, 61
Azteca muelleri, 107

Badgers, 32–33
Bees, 19, 20, 25, 42–43, 44–45, 47, 48, 50, 51, 55, 60, 75, 79, 102, 114, 115, 118, 120–121, 123, 125–127, 129, 130, 143, 144, 147, 149, 151, 153, 160, 161, 165, 166, 167, 168, 171, 173, 174, 175–179
Bellicositermes natalensis. See Termites
Bernard, 14, 128, 178
Birds, 69, 73
Bivouac, of ants, 74–76
Blatella (cockroach), 73, 128, 182–185, 189
Bodenheimer, 26
Boron, 16
Brachinus beetle ("bombardier"), 159
Brains, of ants, 163–164, 201
Brian, 119, 155, 156, 157
Brown and Wilson, 21
Bünzli, 100–101
Butterflies, 59

Camponotinae, 23
  Camponotus, 15, 23, 106, 130–

132, 140, 148, 176, 196
Cardiocondyla, 148
Caterpillars, 28, 96–97, 159, 184
Cavill and Robertson, 161
Cetonia, 32–33
Chauvin, Rémy:
  apparatus of, 56, 63, 67, 186, 187, 190, 191
Chen, 127
Clythra beetle, 33
Coccids. See Scale insects
Cockroaches. See *Blatella*
Collembola, 20, 21, 128
Colobopsis, 15, 106–107, 135
Copulation, in ants, 21, 30
Corpora allata, in ants, 124
Crematogaster. See *Myrmicinae*
Crickets, 111
Cybernetics, 181, 198–199, 201, 202, 204
Cyphomyrmex, 86

Dacetinae, 20
  Daceton armigerum, 21
Darchen, 42
Dendrolasius, 86, 138, 188
  Dendrolasius fuligiosus (black wood ant), 107–108
Dimorphomyrmex, 23
Dinoponera. See *Ponerinae*
Direction-finding, in ants, 171–172
Dobrzanska, 123, 137
Dobrzanski, 137, 179
Dolichoderinae, 14, 23, 97, 100, 135, 154, 159, 160, 176
  Aneuretus, 14
  Dolichoderus gibbifer, 100
Dorylinae (driver ants), 13, 15, 23, 70, 71, 113, 159
  Annoma, 13, 20, 22, 73, 74, 77, 196
  Annoma wilwerthi, 16, 70, 71, 73, 74–75
  Dorylus helvolus, 17
  Eciton, 71, 73
  Eciton burchelli, 71, 73, 74–75
  Eciton hamatum, 72
Dorymyrmex, 88

Driver ants. See *Dorylinae*
Dufour gland, 160–161
Dufour's gland secretion, 150–153, 154, 160
Duns Scotus, 37
Dzierzon's law, 114, 120

Earwigs, 111
Eciton. See *Dorylinae*
Eckstein, 30
Eibl-Eibesfeld, 83, 84
Elasmosoma, 33
Enemies of ants, 32–34
Epimyrma (a parasite), 141
Exchange of food. See Trophallaxis

Fabre, 9
Fielde, Miss, 136
Fire ant. See *Solenopsis saevissima*
Flanders, 24, 98
Flies, 59, 73, 74, 84, 96
Food, 15, 58–60, 61, 66, 77, 116–118, 132
food use, coefficient of, 66–68
food rejected by ants, 66–68
Foraging, 52–69, 148, 151, 156, 178, 179
Forel, 51, 137, 140, 164, 206, 207
Formic acid, 14–15, 20, 32, 69, 131, 154, 158, 159, 161
Formica polyctena (rofo-pratensis). See *Myrmicinae* (red ants)
Formica rufa. See *Myrmicinae* (red ants)
Formica rufo-pratensis. See *Formica polyctena*
Formicinae, 14–15, 132–133, 158, 159, 160, 179, 207
Formica fusca, 23, 104, 105, 124, 127, 135–136, 145–147, 156
Formica neorufibarbis, 138, 205–206
Formica nigricans, 52, 118, 156
Formica sanguinea, 15, 123, 127, 135–136, 137, 141, 142, 147, 206, 207

Formica subnitens, 61
Formica wheeleri, 138–139, 205
Formicoxenus nitidulus, 33
Formiculture, 31–32
Foxes, 32–33
Fungus cultivation, amongst ants, 79–87, 101
Fungus-gardens, 14, 80, 81, 82–83, 85, 86, 128, 135

Geomenotaxy, 173–174
Glands, in ants, 18–20, 123–124, 160–161
Gœtsch, 12, 87, 88, 125, 140, 149–150, 156, 192
Gösswald, 27, 28, 29, 31, 32, 58, 116, 118, 139, 141, 148, 150, 205
Grassé, 40–42
Grylloblatta, 45

Harvester ants. See *Myrmicinae* (*Messor*)
Hays, the, 90
Heating, among ants, 44–52, 75, 81, 115
Heat regulation, 48–50, 52, 81–82
Heer, 23
Herzig, 93, 94, 96, 99
Hester, 46–47
Heymann, 49
Hölldobler, 132
Honeydew, and ants, 15, 92–98, 102, 126, 156, 178
and other insects, 96
Huber, 105
Hymenoptera, parasitic, 24

Ichneumonidae, 69
Infrabuccal sac, 16–18, 84, 101
Insecticides, 90–91, 131
Iridomyrmex humilis (Argentine ant), 97, 162, 196, 206, 207
Iridomyrmex myrmecodiae, 107

Jacoby, 10, 78, 81, 147
Jander, 167, 168, 171
Janet, 131

Kacher, 83
Kalmus, 189
Käthner, 140
Kloft, 93, 94, 116, 148
Kneitz, 47, 49, 51, 209–210
Kül, 121

Labial gland, 18, 19, 116–118, 160
Ladybirds, 59–60
  Cryptolaemus montrouzieri, 103
Landois, 176
Lange, 61, 68
Larvae, of ants, 27, 77, 109, 112–113, 116–119, 150
Lasius, 114, 153, 155, 162, 192
  Lasius brunneus, 100
  Lasius flavus, 99–100, 149, 176
  - Lasius fuliginosus, 88, 205
  Lasius niger, 23, 60, 88, 93, 105, 158, 167
  Lasius umbratus, 131, 138, 205
Ledoux, 110, 120
Le Masne, 129–130, 131
Leptogenys, 20
Leptothorax, 106, 135, 156
Lhoste, 111
Lincecum, 87
Lindauer, Professor, 120, 173
Linsenmaier, 140, 141
Lomechusa beetle, 134, 141–142, 197
Lubbock, 99, 124, 176, 192
Lüscher, 81
Lygaeonematus abietum, 61

MacCook, 206, 207
Majority organ, 200–202
Mandibular glands, 18, 19, 124, 160, 162, 177
Marey, 63

Marikovsky, 30
Markl, 173, 174–175, 177
Maschwitz, 150, 153, 154, 158
Maxillary gland, 18, 19, 123, 160
Megalomyrmex, 135
Megaponera. See Ponerinae
Mermis worm, 33, 129
Messor (harvester ant). See Myrmicinae
Metanotal gland, 16, 18
Meyer, Jean, 54–55, 200–201, 203
Migration, in ants, 70–73, 76
Monomorium pharaonis, 144
Montagner, 144–145
Moths, 59
Muir, 99
Müller, 101
Myopias, 20
Myrmecia, 112, 161, 165, 196
Myrmecia gulosa, 13, 156
Myrmecodia (Javanese plant), 107, 108
Myrmecologists, 9, 10, 34
Myrmicinae, 13–14, 15, 20, 23, 159
  Crematogaster, 14, 86, 108
  Formica lugubris, 181
  Formica polyctena, 29, 46, 49, 60, 68, 88, 95, 103, 104, 105, 133, 114, 118, 125, 127, 154, 156, 179, 181, 186, 187, 190, 206, 208, 209
  Formica rufa, 28–29, 30, 60, 95, 105, 117, 118, 154, 207, 209,
  Leptothorax, 13–14
  Messor (harvester ants), 14, 87–88, 104, 125
  Myrmica, 119, 140, 148, 154, 156, 157, 171, 175
  Myrmica rubra, 53, 88, 144, 157
  Myrmica scabrinodis, 126, 127
  Myrmica schenki, 176
Red ants, 13, 14, 15, 26–69, 94–95, 101, 104–107, 111, 112, 113–115, 118, 119, 121, 123, 124, 132–134, 143, 154, 156, 157, 159, 161, 163–164, 166, 167–171, 173, 178, 185, 191, 196, 197, 206–210

Nachtwey, 176
Neivamyrmex, 79
Nemobius sylvestris (wood grass-hopper), 59
Nests. See Ant nests
Nuptial flight of queen ants, 30, 84–85, 100–101, 113, 139, 162
Nymph-collecting, 27–28

Œcophylla (weaver ants), 15, 73, 98, 102–103, 109–111, 157–158
Œkland, 95, 121
Ommatidia, of ants, 166, 170–171
Orientation, 172, 180–190
Otto, 36, 69, 114, 121–124, 126, 150, 185
Ovaries, of ants, 113, 123

Paltothyreus. See Ponerinae
Parasites, of ants, 128–137, 140–142
Paussus, 129–130
Pavan, 18, 95, 106, 155, 162, 193
Pavan gland, 160, 206
Pest control, by ants, 26, 28, 31–32, 58, 60, 68–69, 101–102, 107
  hampered by ants, 97–98, 102–103
Pharyngeal glands, 124, 160
Pheidole, 87, 88, 98, 125, 144, 162
  Pheidole instabilis, 163–164
Pheidologeton diversus, 15
"Photohorotaxy", 189–190
Phototaxy, 172
Pogonomyrmex, 87, 105, 177
Poison, 14–15, 19–20, 32, 159–162
Poison gland, 19, 160–161
Polyergus rufescens, 15, 135, 136–137, 138, 139, 141
Polyethism (variability of disposition), 125–127
Polymorphism, in ants, 15, 23–24, 74, 79, 80, 83–84
Polyrachis, 109
Ponera eduardi. See Ponerinae
Ponera coarctata. See Ponerinae
Ponerinae, 12, 23, 112, 155, 159, 176
  Megaponera, 12, 176

Paltothyreus, 12, 16, 196
Dinoponera, 12, 196
Ponera eduardi, 22
Ponera coarctata, 23
Pontin, 97
Postpharyngeal gland, 19
Praying mantis, 30, 130
Prenolepsis, 101, 105
Prey, kinds of, hunted by ants, 58–60, 61, 68, 69
Promyrmicides, 13
Proventricular valve, 16
Pseudomyrma, 13, 19, 107

Queen ants, 30–31, 45, 47, 48, 76–79, 84, 111, 113–120, 123, 126, 139–140, 149, 155, 164, 209
Acropyga, 101
Annoma, 13, 20, 22, 74
Atta, 18–19, 20, 79, 84, 101, 177
Camponotus, 22
Dorylinae, 15, 17
Eciton, 74–75, 79
Formica polyctena, 46
Lasius, 60
Myrmica, 119
Solenopsis saevissima, 91

Rabaud, 53, 197–198, 203
Raignier, Father, 13, 22, 30, 49, 50, 70, 73, 75, 77, 112, 176, 196, 206, 207, 209
Rat, white, 180–182
Red ants. See Myrmicinae
Renner, 172, 178
Reservoir-ants, 45
Reyne, 100
Rœder, 30
Roo, 21
Rösch, 120

Sakagami, 127
Scale insects, 92–94, 96–98, 100–101, 102, 103
Scherba, 158
Schmidt, 175

Schneider, 149
Schneirla, 61, 71, 72, 73, 75, 76,
    179, 181–185, 188
Senses (miscellaneous), in ants, 172–
    179
  balance, 172
  hearing, 175
  time, 178–179
Sericomyrmex, 135
Sexuals, 20–21, 48, 118, 155, 166
Sheffer stroke organ, 200–201
Sight, in ants, 38, 59, 145, 165–171,
    189–190
Slavery, in ants, 135–139
Smell, of ants, 12, 146, 149, 150
Soldier ants, 70, 77, 88, 125. See
    also Workers: majores and
    minores
Solenopsis fugax, 135, 166
Solenopsis saevissima (fire ant), 20,
    89–91, 97, 143, 148, 150–153,
    160, 161, 196, 206, 207
Sound-production, by ants, 176–177
Stäger, 60
Stammer, 30, 206, 207, 209
Stary, 98
Stäzer, 135
Steiger, 88
Steiner, 50
Stereomyrmex, 23
Stigmergy, 40–44, 54
Strongylognathus, 139, 140
Strumigenys ludia, 21
Stumper, 135, 140
Sudd, 53, 54, 78, 105, 125
Super-colonies, 206–210
Symphilids, 74
Szlep, 147

Talbot, 105, 156, 158
Tapinoma, 148, 153
Teleutomyrmex, 140–141
Termites, 13, 14, 20, 40–42, 81–82,
    84, 85, 104, 113, 128, 129
  Bellicositermes natalensis, 40, 79
Territory, in ants, 155–158
Tetramorium caespitum, 20, 88, 139,
    140, 148

Thermo - preferendum   (preferred
    temperature), of ants, 46–48
Tools, 109
Torossian, 130–132
Track - making   secretion.   See
    Dufour's gland secretion
Transport of prey, 52–55, 111, 123,
    201
Trophallaxis (exchange of food), 16,
    18, 112, 141, 145–147, 149–
    150, 202

Van de Goot, 97
von Boven, 75
von Frisch, 121, 164–166, 171, 174
von Neumann, 198, 199, 202, 204
Voss, 167–170
Vowles, 168, 175

Wallis, Mrs., 126–127, 145–147
Wasmann, 74, 129, 139, 140, 206,
    207
Wasps, 25, 90, 130, 145, 165
  Polybiinae, 90
  Encyrtus, 97
Way, 102
Weather, effect of upon hunting,
    61–62
Weaver ants. See Œcophylla
Weber, 82, 86
Wellenstein, 59, 65, 69, 144, 172,
    185
Wheeler, 11, 87, 88, 92, 93, 129,
    131, 135, 158, 163, 164
Wilson, 90, 91, 137, 150–153, 155
Woodpeckers, 32
Workers, 20–22, 44–46, 60, 106–
    107, 112–116, 119–120, 120–
    127, 149, 164
  exterior workers, 45–46, 121
  interior workers, 46, 121, 203
  majores and minores (character-
    istics), 120, 125, 148
Wran, 105

Zahn, 45, 47, 48, 51, 150
Zoebelein, 95